Advance Praise for THE HUMAN FACTOR

"What form of social change could save lives, boost the economy, and increase health and happiness, all without political wrangling or moralistic finger-pointing? The answer: making our technology work better with human minds and bodies. This delightful and important book explains how we can at last reap the fruits of the recent revolution in technology."

—STEVEN PINKER, JOHNSTONE PROFESSOR OF PSYCHOLOGY, HARVARD UNIVERSITY, AND AUTHOR OF *The Blank Slate* AND *How the Mind Works*

"This book may well be a landmark in changing our view of technology, and its place in the world. Kim Vicente is a visionary. He places human needs and values first, rather than technology for the sake of technology. The world today badly needs such people."

—ALAN P. LIGHTMAN, AUTHOR OF *Einstein's Dreams*

"Kim Vicente is an engineer who understands how all our lives are being engineered. You will put down this book with a new awareness of the link between devices and those who use them. And you will have been greatly entertained."

—JOHN POLANYI, NOBEL LAUREATE

D1446218

The Human

Factor

revolutionizing the way people live with technology

KIM VICENTE

ROUTLEDGE
New York & London

Published in 2006 by
Routledge
Taylor & Francis Group
270 Madison Avenue
New York, NY 10016

Published in Great Britain by
Routledge
Taylor & Francis Group
2 Park Square
Milton Park, Abingdon
Oxon OX14 4RN

© 2006 by Taylor & Francis Group, LLC
Routledge is an imprint of Taylor & Francis Group
Published in arrangement with Alfred A. Knopf Canada, a division of Random House of Canada, Limited.

Printed in the United States of America on acid-free paper
10 9 8 7 6 5 4 3 2 1

International Standard Book Number-10: 0-415-97891-2 (Softcover)
International Standard Book Number-13: 978-0-415-97891-0 (Softcover)
Library of Congress Card Number 2003067247

No part of this book may be reprinted, reproduced, transmitted, or utilized in any form by any electronic, mechanical, or other means, now known or hereafter invented, including photocopying, microfilming, and recording, or in any information storage or retrieval system, without written permission from the publishers.

Trademark Notice: Product or corporate names may be trademarks or registered trademarks, and are used only for identification and explanation without intent to infringe.

Library of Congress Cataloging-in-Publication Data

Vicente, Kim J.
    The human factor : revolutionizing the way people live with technology / Kim
Vicente.—1st ed.
      p. cm.
  ISBN 0-415-97891-2 (pb : alk. paper)
  1. Technology—Social aspects. 2. Human engineering. 3. Human-computer interaction.
  T14.5 V52 2004
  620.8'2—dc22   200367247

Taylor & Francis Group
is the Academic Division of Informa plc.

**Visit the Taylor & Francis Web site at
http://www.taylorandfrancis.com**

**and the Routledge Web site at
http://www.routledge-ny.com**

To "Dionisio," for setting the example

Contents

Preface   1

Part One
TECHNOLOGY WREAKING HAVOC

1   A Threat to Our Quality of Life: Technology Beyond Our Control   9
2   Why Is Technology So Out of Control? Walking Around Half-Blind   29

Part Two
TECHNOLOGY FOR HUMANS

3   Let's Get Physical: Fitting the Design to the Body   65
4   Minding the Mind I: Everyday Psychology   89
5   Minding the Mind II: Safety-Critical Psychology   111
6   Staying on the Same Page: Choreographing Team Coordination   155
7   Management Matters: Building Learning Organizations   183
8   Political Imperatives I: Technology for Better or for Worse?   231
9   Political Imperatives II: Safeguarding the Public Interest   245

Part Three
REGAINING CONTROL OF OUR LIVES

10 The Way Forward: Not by Widgets Alone   281

Notes   307
Acknowledgments   331
Index   335

Your brain may give birth to any technology, but other brains will decide whether the technology thrives. The number of possible technologies is infinite, and only a few pass this test of affinity with human nature.

ROBERT WRIGHT,
Nonzero: The Logic of Human Destiny

Some of the greatest discoveries . . . consist mainly in the clearing away of psychological roadblocks which obstruct the approach to reality; which is why, *post factum,* they appear so obvious.

ARTHUR KOESTLER,
The Sleepwalkers

Prediction is difficult, especially of the future.

YOGI BERRA

Preface

I realized that my life would never be the same again late one night in the fall of 1983 when I was a third-year undergraduate student in the Department of Industrial Engineering at the University of Toronto. I lived with my parents, and I had taken over their big teak dining-room table to do my homework. My textbooks, papers and class notes were spread out over its entire surface. It was 2:00A.M. and I had been working on a project nonstop since dinner – something not uncommon for an undergraduate engineering student, but this time I noticed an important difference. I had been working hard, but the time had passed quickly, and although it was late, I wasn't tired and wanted to continue working. I realized I was staying up not because I *had* to, but because I *wanted* to. I was fascinated by what I was doing; I was actually enjoying it! I didn't feel that strongly about any other course – differential equations, for instance. There was something different about this course, *human factors engineering* – the unique area of engineering that tailors the design of technology to people, rather than expecting people to adapt to technology.[1] I remember making a mental note to myself – this human factors stuff is pretty cool.

1

Years later, I went on to do a master's and a doctorate, and became a professor of human factors engineering. In retrospect, I can see why the field exerted such a strong attraction: when I was growing up, I did well in science, computer and mathematics courses; I was curious – probably annoyingly so – always trying to figure out how things worked. But I also did relatively well in English classes, loved to read novels, liked being around people, and had a streak of altruism and dedication to socially relevant issues. So on the one hand, I exhibited the stereotypical "geek" profile of a budding engineer, but on the other, I also embraced some of the "artsy" characteristics of an activist and humanist. Human factors engineering, because it deals with people *and* technology, allowed me to pursue all of my interests – to use both sides of my brain, not just half of it.

When I first thought of writing this book, my goal was to explain the social relevance of my discipline to an educated lay audience, because hardly anybody has heard of human factors engineering even though the lack of fit between us and our technologies burdens our lives daily and has even changed the course of human history (I am not exaggerating, as you will see in the pages that follow). But I wound up doing something quite different, and the reasons why are instructive.

About 95 per cent of the work done by human factors engineers is relatively narrow and deals with designing for individual needs: ergonomic office chairs, user-friendly software, and the like. But what I'm going to describe in this book deals with a vastly broader set of problems arising out of the relationship between people and technology, not just at the level of the individual but also at the team level, the organizational and even the political level. I could tell you that this is what human factors engineering is and you'd probably take my word for it because you've never heard of the discipline anyway. But most of the conceptual territory that I'm staking out in this book has *not* been extensively travelled, let alone actively explored or thoroughly mined by my discipline.

That's an important observation because the negative impact of technology on contemporary society goes well beyond the frustrations caused by the myriad user-unfriendly widgets that surround us in the modern world. The negative impact is also clear in much larger problems, like the terrifying impact of fatal medical errors, the irreversible devastation of our natural environment, the deadly threats to safety in the aviation and nuclear power sectors, the contamination of our drinking water, and even the integrity of the democratic process. Don't get me wrong – this isn't a book of complaints and whining about the negative consequences of technology. On the contrary, most of the pages that follow will be devoted to discovering how to make technological systems serve our needs. I hope it will take readers on a journey of understanding, and I'll begin the journey by giving a sense of *why* technology is wreaking havoc, and providing a new way of thinking that makes the human factor central to the design of effective technology in the modern world, whether it be a gadget or a more complex system. Then I'll go on to describe a number of solutions to show what we can and must do to regain control of technology, in our daily lives, in our businesses and in our society at large.

We already know how to design technology that works for people. But what is now becoming clear is that we could apply this knowledge much more widely, we could help solve many persistent social problems of local and global interest and improve the quality of life of everyone on the planet. As a result, the relationship between people and technology isn't just of primary concern to human factors engineers; it's also relevant to a growing number of people from many different walks of life – perhaps you're one of them – people who don't think of themselves as human factors engineers. The problems we're facing as a society are so complex and so pressing that they simply will not yield to any one discipline or profession. They require the adaptation of technology to human nature on a grand scale – what I am calling in this book a "Human-tech Revolution."

As you'll see, the contributions required to achieve a Human-tech Revolution will come not just from human factors engineers, but from other engineers and scientists, as well as from doctors, lawyers, nurses, pilots, administrators, civil servants, executives, environmentalists, educators, managers, politicians, NGO members, sociologists, pharmacists, journalists and Joe and Jane Public. It will take a broad social effort to pull off this revolution. People in many different fields and disciplines have key roles to play in the Human-tech Revolution – all of them bring something important and particular to the table. For instance, some of the ideas I discuss in this book grow out of similar ideas already developed by disciplines other than my own – ideas like learning organizations and risk management programs. If we're to make a world that envisions a happy, successful marriage between technology and human nature, then we should embrace ideas from all quarters because we need all the creative help we can get.

So as I worked on the book, the research I had been doing for the last twenty or so years across many sectors – from nuclear safety to patient safety, for example – and the striking similarities I discovered in the new ideas and approaches being tested in these extremely diverse fields, caused me to rethink my original goals: it no longer seemed enough to explain merely what human factors engineering is. And it became apparent that before we can take a quantum leap forward in designing technology with a human face, we need a common vision and framework to bring all these people and disciplines together to address an essential goal – to use technology to solve our very pressing human and societal problems. Yes, there's a growing amount of activity – perhaps reaching the level of a modest buzz – but it's all happening in different places with little coordination. There's an urgent need to bring all these efforts together synergistically to maximize their impact.

Once I figured this out, I realized that I was no longer writing a book just about my little, unheard-of discipline. Instead, I was writing a book about a new world view, a way of thinking about the role of technology in society that was much broader than what my discipline has achieved to date, a way of thinking that could – actually, must – encompass many diverse disciplines and professions. I now truly believe – more strongly than ever before – that a Human-tech Revolution can provide the unifying goal and conceptual framework we need to unite all the required strands of effort and expertise. This new world view offers a more sophisticated, elegant way of thinking that can help us harness the power of technology to society's benefit. And as I'll show in this book, lack of ability or understanding is *not* the primary stumbling block to advancing the revolution. We already have more than enough knowledge to do far better than we're currently doing. Yes, there are many questions that remain to be investigated; but if we concentrate on the human factor and make it central to the technological world in which we now live, that world could be a completely different place – safer, healthier, more productive and sustainable, and more humane.

Part One

TECHNOLOGY WREAKING HAVOC

1

A Threat to Our Quality of Life:
Technology Beyond Our Control

Just before midnight on Friday, April 25, 1986, Leonid Toptunov was about to begin the graveyard shift in the control room of the Vladimir Ilyich Lenin nuclear power station located near Chernobyl, just 130 kilometres northeast of Kiev and 600 kilometres southwest of Moscow.[1] The weather had been unseasonably warm that week, but the joyous May Day holiday celebrations were less than a week away. As Toptunov took off his street clothes and donned his pristine white overalls and white beret for the last time, he had no idea that in less than two hours he would become an unwitting participant in a catastrophic event of historic proportions.

Earlier that day, the Chernobyl operators had begun an experimental test. It required two important conditions be satisfied: the power being produced by the nuclear reactor had to be reduced to about 25 per cent of its full capacity, and the primary safety system that was designed to protect the plant during an emergency had to be turned off – during the entire period of the test. At 1:00 P.M. the operators had begun to reduce the amount of power produced by the nuclear reactor, closely monitoring the relevant meters on the immense technological consoles in front of them. One hour later, they deliberately disabled the safety system, stripping the plant of one of its primary defences – all as required by the test plan. A nine-hour delay ensued. The continuation of the test was put off until a later shift.

Nuclear reactors have very complex dynamics, and Chernobyl was no exception. As a result of this complexity, Toptunov – the senior reactor control engineer on his crew – had trouble stopping the power level at 25 per cent and actually bottomed out at a power level of 7 per cent. But the Soviet RBMK-1000 reactor design is very unstable at low power, making it exceedingly difficult for operators to maintain control of the plant. This, combined with the fact that one of the main safety systems was turned off, made the situation extremely dangerous, but Toptunov and his comrades didn't realize it because they weren't used to running the reactor at such a low power level and because they didn't fully understand the complex principles governing the reactor's behaviour. To make matters worse, the thousands of indicators on the wall-sized consoles in front of Toptunov presented a bewildering array of *data,* but not enough *information,* and so the gravity of the situation wasn't obvious to him. And besides, young Toptunov had been told that technical experts had estimated the likelihood of a severe accident to be one in ten million – a virtual impossibility. So he and his co-workers persisted with the test.

To do so they improvised – with the plant in an unfamiliar and increasingly dangerous state – eventually stripping the reactor of its remaining safety systems. By 1:22 A.M. that fateful night, the nuclear reactor was almost out of control. Yet the temperature in the control room didn't skyrocket, there was no growing vibration, no loud noises – nothing comparable to what was shortly to come. The only thing that changed was the set of indications on the displays embedded in the bewildering consoles. Just two minutes later, at 1:24 A.M., Toptunov finally realized that the readings staring him in the face meant that a terrible event

was about to occur: in a last-ditch effort to avert disaster, he tried
to turn off the reactor. But his well-intentioned effort came too
late; by that point, Chernobyl's fate was sealed. A critical nuclear
reaction – the type that occurs by design in an atomic bomb but
is never supposed to happen in a nuclear power plant – was
inevitable. And immediate.

The first violent explosion unleashed a power spike one hundred times greater than anything the reactor was designed to
produce under normal operating conditions. It hoisted the
thousand-ton steel and concrete plate covering the reactor,
exposing the 1,680 nuclear fuel rods in the reactor core and
spewing deadly radioactivity into the atmosphere. The force of
the explosion was so mighty that it sent radioactive particles flying as high as one kilometre into the air. A second furious explosion caused the graphite in the reactor core to burst into flames.
The fire continued to burn for nine days, releasing an invisible,
constant stream of radioactive particles into the environment. The
reactor itself was destroyed.

Until that instant, when the first explosion ripped the reactor apart, the nuclear technology had functioned precisely as
intended. The designers had done everything they were supposed to do from a technical perspective: all the hardware and
software worked flawlessly. And Toptunov and his colleagues
were carrying out the test plan as well as they knew how. The
problem was that the plant designers hadn't paid enough
attention to *the human factor* – the operators were trained but
the complexity of the reactor and the control panels nevertheless outstripped their ability to grasp what they were seeing.[2] Toptunov didn't completely understand the effects his
actions were going to have until it was too late – with devastating consequences. When the graphite reactor core exploded
into flames, the awesome impact that a nuclear power plant
can have on both humankind and the environment was
vividly realized.[3]

The six hundred people unlucky enough to be working at the plant that evening received very high doses of radiation and many later suffered lingering or fatal diseases. The 116,000 people who were evacuated from the neighbouring farms and towns received lower but still significant doses of radiation. The 600,000 military and civilian workers who heroically helped put out the fires, evacuate the public, and clean up the disaster were also exposed to high levels of radiation. About 140 people experienced various degrees of injuries, including convulsive radiation sickness and burns that caused blistered skin to slide off the flesh. A total of thirty-one people died as a result of the accident, including Toptunov.

He was twenty-six years old.

One of the further horrors of a nuclear catastrophe is that its impact travels widely across time and space. The number of cases of thyroid cancer among children in the area has increased, with almost 1,800 diagnosed between 1990 and 1998. Harder to measure, but just as real, is the psychological impact of such a disaster: one of the most significant health effects of the Chernobyl accident was the mental anguish and trauma experienced by the local population. People continue to be terrified of the unknown effects of radiation; they don't trust the government or scientific experts, and their way of life has been severely disrupted. These health effects will persist for generations.

But the environmental contamination is equally lasting because there is no "undo" command for a nuclear accident. To this day, large areas of land can no longer be used for agricultural purposes, and food is still monitored for radiation over an even larger area. And the impact of a nuclear accident on this scale transcends geographical borders. Chernobyl released radioactive material all across the northern hemisphere, although Europe was hardest hit. The degree of contamination outside the Soviet Union was relatively low, but radioactive fallout was detected and measured in England, Scandinavia, Southern Europe, Canada, the United States, and as far away as

Japan, with the exact amount depending on the weather – if there was rainfall in a particular area when the radioactive cloud passed over, it received a greater amount of radioactivity. The lesson became clear with Chernobyl – a nuclear catastrophe anywhere can be a nuclear catastrophe everywhere.

Step back a moment to 1936. In the final days of black-and-white silent movies, Charlie Chaplin created a masterful satire of industrialization, *Modern Times,* which drew attention to the human and social costs of technology. In one memorable sequence, Chaplin is shown working on an assembly line. His job is to perform a few motions over and over again; he uses two wrenches, one in each hand, to tighten two bolts on each component rolling by on a conveyor belt. The speed of the belt increases; Chaplin tries desperately to keep up but eventually he's carried away by the conveyor belt and fed into a chute. In the next scene, we see several gigantic mechanical wheels with intertwined geared teeth grinding the bemused Little Tramp through a rigidly defined S-shaped path, first forward, then back, then forward again. He has been forced to adapt to technology – literally: he has become a veritable cog in the wheel.

Chaplin, however, had to adapt only to mechanical gears moving at terrestrial speed. We who inhabit the modern times of the twenty-first century have to adapt to digital technology moving at light speed. More and more technology is being foisted upon us at a faster and faster pace. We walk around with electronic leashes – pagers, cell phones, personal digital assistants and pocket PCs – that tie us to our work. At home, we have the latest electronic consumer products – each with its own remote control and hefty user's manual. All these gadgets are supposed to make life easier, but they often make it more difficult instead. And before we learn to use the latest technological "convenience," there's a new one on the market with more "advanced" features.

No matter how many user's manuals we read, we just can't seem to keep up.

The challenges facing us have never been more daunting, despite the fact that our knowledge of the physical world and the technological possibilities we possess are vastly more sophisticated than they were even fifty years ago. Never before in the history of human civilization have we so quickly amassed so much knowledge of science, mathematics and engineering, and never before have we seen such tremendous advances in technology. The number, diversity and sophistication of the options available to us allow us to conceive and construct increasingly intricate products and systems. Given this abundant knowledge of both the physical world and of technological possibilities, we might expect our problems with technology to decrease, not increase. Granted, many technical innovations have undoubtedly improved our quality of life. One well-known example is the PalmPilot personal digital assistant. This hand-held electronic device has been a marketplace success because many people find it both useful and easy to use. In later chapters, I'll describe how the PalmPilot and several other successful everyday products were designed. But devices that are easy for people to use and that serve a significant human or societal need seem to be the exception. As a result, there's a growing realization that all is not well in the world of technology.[4]

Here's an everyday example. A few years ago, Mercedes-Benz started offering a feature on their E320 model that lets drivers check their oil electronically, from the driver's seat.[5] It seems like a clever use of technology. You don't have to leave the cosy confines of your climate-controlled automobile. Smart. You no longer have to pop the hood, find a rag to wipe the dipstick, or figure out which of the several dipstick-looking things under the hood is really the dipstick. And you don't have to go through the tedious and messy manual process of lifting the dipstick, wiping it, reinserting it, taking a reading and reinserting it again –

exactly the kind of innovation you'd expect from legendary German engineering.

This electronic oil-checking feature couldn't have been designed decades ago, before the transistor was invented. At that time, our knowledge of electronics and our available technological options were too impoverished to permit such a potentially useful feature. I say "potentially," because I haven't yet described what you actually have to do to check your oil from the driver's seat in this car. There are only five steps. Step number 1: turn the car off. Step number 2: wait for the oil to settle. Fair enough. It doesn't make sense to check the oil with the engine on. You have to let things settle to get a reliable reading of the level. Step number 3: turn the ignition two notches to the right. Hmmm. That's a little less obvious. It's easy enough to do, but there's no intuitive relationship between the action and the effect of the action. Step number 4: wait five seconds. What? Wait five seconds? You've already waited for the oil to settle. Why do you have to wait another five seconds? But you're not done yet. There's one more step. Step number 5: within one second, press the odometer reset button twice. This step makes no sense whatsoever. It seems completely arbitrary. What does the odometer reset button have to do with checking the oil? As far as I can tell, there's no logical answer to this question – and I have a Ph.D. in mechanical engineering. The average driver will be baffled, even though the electronic components have been painstakingly designed, with a sophisticated understanding of the laws of electricity. In the end, most people will just get out of the car and check the oil the old-fashioned way because they can't remember the steps and can't be bothered to read the counterintuitive instructions again. So much for that legendary German engineering.

I once tried to describe the work my students and I do to a journalist, who turned my long-winded explanation into a succinct sound bite: "Oh, so you're technological anthropologists!" I had never thought of our work in that way, but I suppose that's one way of describing it. We have indeed done a number of field studies of people using technology *in situ* – or in their local habitat, my journalist friend might say. I once spent an entire week during spring break on the twelve-hour night shift in a nuclear power plant's control room trying to figure out how the operators performed what looked like an impossible job, incredibly reliably, day in and day out. I've also spent time in hospitals, just talking to doctors and nurses about how technology helps or hinders their jobs, and watching surgeries in the O.R. One of the first operations I saw was an amputation below the knee (my medical colleagues didn't inform me ahead of time – if they had, I might not have showed up). And more recently, I've spent days in 911 call centres listening in on phone operators and ambulance dispatchers trying to deal with life-and-death medical emergencies. (In one call, the 911 operator was trying to guide the caller to perform mouth-to-mouth resuscitation on an old man who had started to turn blue, but the caller was reluctant to get that close to the ailing patient, giving as an excuse not to follow the instructions: "I think he's dead, I think he's dead!") My graduate students have spent countless hours talking to operators in petrochemical plant control rooms, watching computer network managers monitor and troubleshoot telecommunication webs, sitting in with flight engineers in aircraft simulators and in real cockpits on long flights, observing nurses programming computer-based medical devices in hospital recovery rooms, and observing an engineering design firm at work, over a period of months. At the same time, we've conducted research on how to design better technological systems – generating new design ideas, building prototypes and running controlled scientific experiments to see if our creations really do help people do their jobs better.

As a byproduct of our obsession with the interaction of people and technology, we've served as consultants for large corporations and governments. Some of our ideas have actually been adopted by industry – a rare coup for a bunch of academics! (I remember the first time I saw something I designed being used by a company – a prototype nuclear power plant control room of the future built by Toshiba, in Japan. My first thought was: "Wow, that's really cool! I created that display." My second thought was: "But what if it doesn't work?") And of course, we've written many research proposals, published lots of papers, given presentations around the world, sat on advisory committees and tried to explain to the media why what we do is important. But most of the time, we've just been having a lot of fun. Because watching ordinary people struggle with technology, learning about the ingenious tricks they conjure up to make their jobs more manageable, inventing design ideas to help them do their work more effectively, and then trying to persuade people in the corporate world and in government to adopt the more promising ideas is a truly gratifying and fascinating way to spend time – let alone earn a paycheque. And I happen to think it's also a matter of urgent importance in our modern world – and becoming more so every year.

The experience I've gained in doing this type of work since 1983 has convinced me of one distressing thing that most people – who spend their time in more conventional ways – aren't even aware of. The pattern revealed by the small, everyday example of the oil-checking gizmo is repeating itself in all aspects of our lives. More and more, we're being asked to live with technology that is technically reliable, because it was created to fit our knowledge of the physical world, but that is so complex or so counterintuitive that it's actually unusable by most human beings. Even in the relatively benign context of everyday tasks, this pattern is already creating dysfunctional effects. It leads to human error, anger and frustration; we've all felt our blood pressure rise when we're lost in the labyrinth of options offered by automated phone message systems or when we're trying to guess which light switch corresponds to the set of lights we want to turn on or off.

Eventually these inefficiencies, errors and maddeningly complex situations give way to alienation and in the long run this leads to an even more severe double whammy: a failure to exploit the potential of both people *and* technology. Human beings are capable of doing some pretty remarkable things, but if we become alienated from technology our full capacities won't be realized. Great technological innovations will go underutilized, and enormous corporate investments in technological development and deployment will go up in smoke.

And when we add up all the negative effects, we can see that our everyday difficulties with technology don't just create problems for us individually; they're also causing society a raft of problems – psychosocial difficulties, loss of productivity, economic distress and more – problems we can't afford.[6] The impact on our quality of life is disconcerting.

Unfortunately, this pattern – technology that is well tailored to the physical world but too complex for human beings to handle – isn't restricted to everyday gadgets like electronic oil checkers; it's also found in larger, safety-critical technological sectors. And there the dysfunctional effects of complexity can be lethal. Nor do we have to look so far afield as Chernobyl to see this clearly. Bad as that was, the threat posed to our lives by such rare catastrophic events is actually dwarfed by a peril that is so devastating, yet so unnoticed, that it has been referred to as a "hidden epidemic." On November 29, 1999, the U.S. Institute of Medicine (IOM) – a distinguished medical body that provides public policy advice to Congress – released a landmark report documenting the deadly impact of medical error on patient safety in the United States.[7] Those of us who had conducted research on medical error already knew the statistics, but the general public didn't. The bombshell was that human error in medicine was conservatively estimated to account for between 44,000 and 98,000 *preventable* hospital deaths annually in the United States alone.[8]

These estimates are so large that they're difficult for us to really understand in terms of everyday experience, but perhaps a few comparisons will help. If the preventable mortality rate were the same in commercial aviation as it is in health care, then a wide-body jet-aircraft accident with no survivors would occur *once every day or two*. If you take the conservative lower estimate of 44,000 preventable deaths, then medical error is the eighth leading cause of death in the United States. It kills more people than AIDS (16,516), breast cancer (42,297) and even traffic accidents (43,458). The annual cost of preventable errors resulting in patient injury has been estimated to be between US $19 billion and $26 billion. And don't forget, these estimates are for the United States – the paragon of health care research and high-tech medicine – not a poor underdeveloped country with a fragile and immature health care system. So it shouldn't be surprising to find that equally appalling mortality or financial statistics have been found in studies conducted in Australia, New Zealand, Denmark and the UK.[9]

Since the IOM report was released, the statistics it cited have been a subject of heated debate in the medical literature. Some researchers say the estimates are too high, others say they probably underestimate the real magnitude of the problem.[10] But no one denies that too many preventable deaths do occur, and for the average person on the street, this academic debate is irrelevant; it doesn't matter whether the true number of people who get killed annually in the U.S. from preventable medical error is "only" 10,000 or over 100,000. Either scenario is a scandal and is completely unacceptable. Which is why the respected IOM didn't pull any punches in making its point:

> The status quo is not acceptable and cannot be tolerated any longer. Despite the cost pressures, liability constraints, resistance

to change and other seemingly insurmountable barriers, it is simply not acceptable for patients to be harmed by the same health care system that is supposed to offer healing and comfort. "First do no harm" is an often quoted term from Hippocrates. Everyone working in health care is familiar with the term. At a very minimum, the health system needs to offer that assurance and security to the public.[11]

There isn't a sliver of ambiguity in this message. Hippocrates must be turning in his grave.

The problem in health care isn't just difficult-to-use hardware and software – although there's certainly plenty of that – but also a work environment that doesn't appreciate what we know about human beings. So far, I've been using the word "technology" in its most familiar sense, referring to materials and their configuration; in short, the physical stuff, such as nuclear power plants, electronic components and so forth. From now on, I'm going to use "technology" in a much broader sense, to include not just the physical but also the non-physical stuff that can be found in complex technological systems (nuclear power plants, public water utilities and the like) – "softer" elements such as work schedules, information, team responsibilities, staffing, and even legal regulations. I have two primary reasons for adopting this broader definition of technology, one a bit pedantic and the other more practical.

First, this broader perspective fits one of the definitions of technology listed by Webster's dictionary: "the system by which a society provides its members with those things needed or desired."[12] Under this big-tent view, any tool – physical, virtual, conceptual or cultural – that helps people make decisions, act and achieve their goals is a technology. All the non-physical examples above – the "soft" elements – fall under this definition. A work schedule, for instance, is part of the system by which a hospital, say, provides its citizens with health care, or by which the aviation industry moves us around the globe. This way of thinking about technology is admittedly unusual, but it has been adopted productively by some people, most notably by the social critic Neil Postman, author of *Technopoly: The Surrender of Culture to Technology*.[13]

My second reason for adopting this broad view is that non-physical things, such as our work schedules, play a crucial role in determining how well a technological system as a whole functions. In other words, it's not just the design of overly complex hardware or software that can cause problems for people, it's also the design of the less tangible aspects of the system. I know of no better way to illustrate this point than to describe for you the impact of hospital work schedules on fatigued medical workers.

That rare commodity, common sense, tells us that we need to sleep to function properly. If we're deprived of sleep for an extended period, we're not "at our best." It's not just that we get cranky – we also find it harder to be attentive; we make more mistakes. That's why there are legal regulations on working hours in so many safety-critical industries. Airline pilots, truck drivers, bus drivers, railroad crews, nuclear power plant operators and maritime crews all have to observe strict limits. For example, American commercial airline pilots are required to have at least eight hours of rest between flight periods and aren't allowed to fly more than thirty hours in one week.[14] New rules have been proposed, reducing these work-hour limits even further, and having such restrictions in place seems eminently reasonable.[15] How confident would you be if you were boarding an eight-hour flight from Toronto to Copenhagen and you learned that your flight crew had just flown from Tokyo to Toronto and hadn't slept in twenty hours? How much faith would you have in that crew, given their physiological state? In medicine, situations like this one are commonplace; they happen every single day. Medical residents often work more than 30 hours in a row, and 100 to 120 hours in a week.[16] (If you haven't counted lately, there are 168 hours in a week.)

In November of 1998, I attended a multidisciplinary conference on how to reduce errors in health care, held at the renowned Annenberg Center for Health Sciences in Rancho Mirage, California. The most disturbing and thought-provoking presentation was on medical fatigue. It was given by Dr. Bertrand Bell, Distinguished University Professor at Albert Einstein College of Medicine at Yeshiva University in the Bronx. He began by describing the tragic death of eighteen-year-old Libby Zion in New York City.

On March 4, 1984, Libby was admitted to Cornell Medical Center's New York Hospital with a high fever. The next morning she was dead. An investigation revealed that an adverse reaction between two medications was the cause of death. However, further digging uncovered another important fact: Libby had been cared for by a sleep-deprived and overworked medical resident (a physician in training). Sidney Zion, Libby's father, believed that if the resident had been more alert she might have realized that the two drugs weren't supposed to be given together. And since Sidney was a well-known journalist, his daughter's death received a great deal of publicity, eventually leading the governor of New York State to appoint a committee to review the case and make recommendations. Dr. Bertrand Bell chaired that committee, and became intimately familiar with the case. Because sleep deprivation and chronic fatigue were implicated as contributing factors in Libby's death, Bell's committee recommended that legislation be passed to reduce the hours that medical residents work. In 1989, four years after her death, these recommendations were signed into law in New York State. Medical residents would not be allowed to work more than twenty-four consecutive hours. Also, their work week couldn't exceed an average of eighty hours over a four-week period. That still sounds like an absurdly heavy work schedule compared to the thirty-hour work-week limits in aviation, but the new regulations were well below the 120-hour work-week practices that were normal in health care.

There was one hitch though. Hospitals complained that it would cost a lot of money to obey these new laws. Who was going to pay for the added expenses? they asked. In response, the New York state government allocated US $240 million per year to help pay for the costs of implementing the new laws.

But in 1998, when the New York State Department of Health conducted surprise inspections at several hospitals, they found that the laws were being routinely violated. For example, 77 per cent of the surgical residents in New York City were still working more than ninety-five hours a week – almost two and a half times the forty-hour work week that is the lot of most people. And even today, the problem has not been solved: on June 26, 2002, the New York State Department of Health announced that 54 of the 82 hospitals that it had inspected since November of 2001 were still violating the legal limits instituted after Libby Zion's death.[17]

Doctors aren't the only ones working insanely long hours in health care across North America. In 1998–1999, the amount of overtime worked by nurses in acute-care hospitals in the province of Ontario was equivalent to 2,250 full-time nurses.[18] This situation is even more appalling when you consider that most of the overtime occurs right after a twelve-hour shift. Furthermore, the amount of overtime is increasing, threatening not just patient safety, but the health of the nurses themselves. No studies have been carried out in Canada to determine how this overtime correlates with hospital errors, but data from the United States suggest that a direct link is likely.[19]

Clearly, safety can be threatened not just when the physical components of a system are too complex for people to understand, as in the case of Chernobyl, but also when non-physical factors – for instance work schedules – affect the performance of the people working within the system. The implication is plain: when we set out to design complex technological systems, we should focus on the physical *and* the non-physical aspects of the system. It may seem odd to think of "designing" a "non-physical technology," such as a work schedule (or even an organizational structure or a piece of legislation). But just as designers choose from all the various possible materials when constructing a bridge, designers also need to select from all the possible work schedules when constructing a health-care system. In either case, the wrong choice can threaten safety. In fact, the non-physical aspects of organizations and industries now play an even greater role than the physical aspects, as I hope to show you. And broadening our view of technology to include both physical and non-physical aspects of system design has great pragmatic value, because it also reveals how we can plan our work environments to embrace the human factor.

Based on my twenty years of travel as a "technological anthro-pologist," I regret to inform you that the problems posed by technology aren't restricted to just one or two sectors. And while I don't have space in this book to cover everything, I intend to draw primarily on examples that affect most of us closely, beginning with the consumer products that we all use in our everyday lives. Then we'll look at more complex systems: aviation, airport security and the environment, along with nuclear power and health care – all "safety-critical" sectors. I chose these particular sectors for three reasons. First, because examples of successes as well as failures can be found in each, although some sectors are clearly in far better shape than others; and second, because they have implications for a much broader range of sectors, such as the petrochemical, aerospace, maritime, automotive, railway and air traffic control industries.

Finally, and most important of all, each of these safety-critical sectors has a dramatic impact on the quality of life of everyone on the planet. Just look at the range and magnitude of the threats to humankind posed by some of them, beginning with the least dangerous (commercial aviation), and then moving to airport security and the environment.

First, commercial aviation: at least for now, this is – comparatively – a remarkably safe complex technological system. There are usually over 10 million commercial aviation takeoffs and landings each year in the United States, yet the accident rate is typically less than one in a million. For instance, from 1984 to 1996, there was an average of nine fatal accidents per year, leading to an average of 204 deaths. Compare that to automobile fatalities: each year in the United States, about 40,000 people die from motor vehicle accidents. Against this backdrop, the aviation figures are downright impressive.[20] And flying wasn't always this safe, as we'll see later.

Nevertheless, there is still grave cause for concern in the near future. Flying plays an increasingly important role in the business and holiday plans of millions of people around the world, so the global amount of air traffic is increasing and will continue to do so. In China in particular, air traffic is expected to grow dramatically over the next few years. So even if the probability of an aviation accident stays at its current impressively low level, by the year 2015 there may be a major plane crash every seven to ten days.[21] That means that about once a fortnight, when we turn on the TV, we'll see live news coverage of another airplane falling out of the sky with the usual terrifying and tragic consequences. Would this affect your willingness to get on an airplane? Statisticians would tell you that you shouldn't pay any attention to the increase in the absolute number of accidents because the probability of an accident is still the same as it was a few years ago – it's just that there are so many more flights now. I doubt whether that kind of number-crunching will truly ease your mind. Unless we take some drastic measures, flight safety is about to slip out of control, because of the vast increase in air traffic in our skies in the immediate future.

Airport security hasn't been in the best shape either. Billions of dollars have been allocated to improving airport security systems since September 11, 2001, and it's difficult to assess their level of safety when so many changes are still being made. However, the effectiveness of aviation security before September 11th is well documented.[22] For example, in a test conducted in 2000, investigators from the U.S. Department of Transportation breached airport security systems 117 times out of 173 attempts – a rather scary security failure rate of 68 per cent. This figure becomes more disturbing when we consider that the United States averages over 1.5 million air passengers each day and the U.K. about 500,000. Given these tremendous volumes, even a very high success rate would still result in many breaches of security. Fortunately, terrorists represent a tiny proportion of the flying population, but here again, statistics may not be entirely comforting.

Then there's perhaps the most complex sector of all: the environment. Ecological degradation is posing a dreadful threat to our quality of life. Statistics from studies of pollution, climate change and species extinction overwhelmingly show the severity and breadth of the global problems we're facing. And by "we," I do mean everyone, because no one is immune to environmental degradation. Here are just a few examples from around the world to justify our collective concern:[23]

- Public drinking water systems are being contaminated with killer bacteria and other lethal waste products, even in developed nations. In the Canadian province of Saskatchewan alone, over 79 boil-water advisories were issued during the latter half of 2001.
- From 1986–88 to 1996–98, the amount of paper and paperboard created from forest production increased tremendously in many countries: 123 per cent in China,

502 per cent in Indonesia, 148 per cent in the Republic of Korea, 104 per cent in Greece, and a whopping 770 per cent in Singapore.

- Global electricity consumption is growing exponentially and about 68 per cent of the energy used worldwide is still generated by fossil fuels. Such fuels produce CO_2 and other greenhouse-effect gases that do long-term damage to the environment.

- An average of nine tonnes of mass is consumed in the United States per person per year, 90 per cent of which becomes waste, thereby creating tremendous pollution problems.

- The global population may reach 10 billion by the year 2025, aggravating the problems I've just listed.

No matter where we look, whether at everyday situations or complex systems, we see technology that's beyond our human capacity to control. In the more mundane cases, like the electronic oil-checking gizmo, the day-to-day results we all experience are bad enough – inefficiency, frustration, alienation and a failure to realize our human and technological potential. But when we turn to safety-critical sectors – nuclear power, health care, aviation, airport security and the environment – the consequences of technology running amok are far more worrisome. Errors in these complex systems can lead to expensive industrial accidents, such as airplane crashes, each costing millions or billions of dollars in damage, not to mention the inestimable cost in human life. Complex technological systems out of control can also lead to expensive litigation, because individuals and organizations frequently get sued when things go wrong. In some cases, errors in these systems can lead to ecological disasters that threaten the environment, such as the contamination caused not only by Chernobyl, but by the enormous *Exxon Valdez* oil spill off the coast of Alaska. These costs are a huge burden on society. And in our connected world, poorly designed complex technological systems endanger all nations, not just developed countries. Even though a large proportion of the global population has never seen a VCR, or any other electronic device for that matter, they can't escape the effects of technology, as Chernobyl made abundantly clear. The industrial world is exporting more and more of its technologies to non-industrialized countries, sometimes without much thought about the impact that those technologies will have in other cultures – witness the chemical plant disaster in Bhopal, India. And ironically, measures to counteract growing fears of global terrorism merely add to the mix. If up to 98,000 Americans die annually from preventable medical error when the U.S. *isn't* being besieged by terrorist threats, just imagine the potential unanticipated safety threats created by the logistical nightmare of having to quickly inoculate an entire nation of 300 million people against smallpox – the most explosive biological weapon on the face of the earth.[24]

Few people are aware of the immense magnitude and breadth of the threat posed by complex technological systems because they haven't learned to see the pattern that links our frustration with overly complex electronic gadgets to the lethal threats posed by medical errors and nuclear meltdowns. But that's what I've been paying close attention to – as have a number of my colleagues and my students. Technology – with all its promise and potential – has gotten so far beyond human control that it's threatening the future of humankind.

2

Why Is Technology So Out of Control?
Walking Around Half-Blind

TWO ANTIQUATED CULTURES IN MODERN TIMES:
THE MECHANISTIC AND HUMANISTIC WORLD VIEWS

Why is technology spinning beyond our control?

There's one explanation we can dismiss outright. Designers don't *deliberately* build uncontrollable technological systems. I haven't talked to the inventors of the electronic oil-checking device I described earlier, but I'm pretty sure they had the best of intentions. They didn't sit around laughing maliciously and saying, "OK, we've made it possible for people to check the oil from the driver's seat. Now let's design a set of steps that's really hard to remember. And, let's make the last step pressing the odometer reset button twice within one second – that'll really drive them crazy!"

No, it's not that simple. The real reasons for our technological woes go back a long way. In fact, to really understand what's going on we have to look at some of the fundamental principles of our approach to the world we live in – the organization of human knowledge. For the last several hundred years, we've adopted a *reductionist* approach to solving problems, deconstructing them into their smallest parts and then studying those in relative isolation. In the eighteenth century, the French mathematician Pierre Simon de Laplace took this slice-and-dice philosophy to an extreme, believing that if we could only break the entire universe down into its elementary particles and account for the motion of those individual particles, we would be able to understand, well, everything. That may seem like a crazy idea now (can quarks explain why people fall in love?), but at that time, Laplace's insight was incredibly powerful, and it wound up having a tremendous impact on the history of ideas.

This general approach also gave rise to another intellectual

The traditional reductionist approach to organizing human knowledge. Each discipline studies the world from a roughly independent perspective. creating a silo effect.

habit, one that is more directly relevant to the concerns of this book: we tend to divide up what we know into categories (or "silos") defined by rigid disciplinary boundaries, such as physics, biology, chemistry, psychology, religion and art. These traditional categories of knowledge made it easier for us to tackle otherwise unmanageable questions. Instead of trying to understand the whole world in its overwhelming detail all at once, we developed a divide-and-conquer approach – you study the electrons, I'll study the neurons, and when we're both done we'll put our respective pieces of the puzzle together to form the big picture. That's the hope, anyway. And it's a way of thinking that has been extremely influential. In the seventeenth century, the French philosopher René Descartes drew a distinction between mind and body that still shapes the way many people think about their disciplines. They pay attention to what falls inside their purview, assuming that anything beyond that can be safely ignored. This approach was enduringly valuable: it led to enormous progress in human thought, from the discovery of the atom to the mapping of the human genome.

But as the English novelist and scientist C. P. Snow pointed out in his classic 1959 essay *The Two Cultures,* specialization has taken a toll. He worried that "the intellectual life of the whole of western society is increasingly being split into two polar groups" – science and art. The gulf between technical/analytical thinking on the one hand and creative/Humanistic thinking on the other was already so deep that "those in the two cultures can't talk to each other." Snow wasn't just referring to uncomfortable silences at cocktail parties. The division he identified had serious consequences: "When those two senses have grown apart, then no society is going to be able to think with wisdom . . . This polarisation is sheer loss to us all. To us as people, and to our society."[1]

Snow couldn't know how true his words would still ring in the twenty-first century. Since his time, matters have only got worse. The reductionist strategy that led to the two-cultures problem has also led directly to our troubles with technology. In this case, I'm referring not to Snow's division between lab-coat scientists with pocket-protectors and vagabond poets wearing black, but rather to a rift within science itself; we've divided scientific knowledge, broadly, into two big groups: the human sciences and the technical sciences. The first group has adopted a *Humanistic* view; when they look at the world, they focus primarily on people. For example, cognitive psychology studies how the human mind works, but it doesn't often consider the mental activity of ordinary people using tools like calculators, cars, computers or appliances to perform everyday tasks; an understanding of technology is missing from the equation.[2] Conversely, the technical sciences – engineering, computer science and applied mathematics – have adopted a *Mechanistic* view; when they

Societal Needs

Humanistic View   Mechanistic View

Traditional disciplinary boundaries create a division between the human and technical sciences, so neither the Humanistic nor the Mechanistic views can clearly see the relationship between people and technology.

look at the world, they focus primarily on the hardware or software; an understanding of human needs and capabilities isn't part of the equation. When computer engineers design tiny gadgets that can process a great deal of information very quickly, they don't think about the characteristics or needs of the people who will be using those gadgets.[3] Until recently, this straightforward division of scientific labour seemed like a reasonable way to make sense of our world.

Unfortunately, this traditional approach has created two breeds of Cyclops – the one-eyed Humanist who can focus on people but not technology, and the one-eyed Mechanistic variety who knows about technology but not people. We're all walking around half-blind. To make matters worse, the Humanistic and Mechanistic world views rarely meet, as anyone who has ever set foot on a university campus knows. There are artsy Humanists and there are geeky technologists, and people tend to be educated to become one or the other.

We're so used to defining people this way that it's easy to forget that the traditional Humanistic and Mechanistic world views are both abstractions of convenience; nobody has ever seen technology without people, or people without technology. In the real world, people and technology co-exist. In fact, the capability to build or use tools is part of what it means to be a human being. Our disciplinary divisions don't represent the world as it really is, with people and technology side by side and interacting.

It bears emphasizing: *our traditional ways of thinking have ignored – and virtually made invisible – the relationship between people and technology.*

But since we think we can choose between these two world views, it makes perfect sense to put people trained in the Cyclopean Mechanistic world view – let's call them the Wizards – in charge of designing technology. After all, they're the ones who can design airplanes, power plants, cell phones, Blackberries and other technical marvels. People trained from a Cyclopean Humanistic perspective don't generally have the technical know-how – the in-depth knowledge of mathematics, physics and computers – to design and build dependable technology. So, generally, the Wizards are put in sole charge of technological development because we don't believe the technically challenged Humanists have anything to contribute. The net result is that technological systems are therefore usually reliable only from a narrow technical perspective – because their designers had the know-how to develop products or systems that have an affinity with the relevant aspects of the physical world, and the physical world only.

Any attempt to explain why technology is spinning out of control must take three other observations into account. The first is that the technical stuff is frequently too complex for people to manage, creating confusion at best and potentially devastating consequences at worst. The second is that the "softer" aspects of technological systems (work schedules, team coordination, and so on) can also make people's lives more difficult than they need be, contributing to the chaos. And thirdly, to top it off, our problems with technology are only getting worse, not better. How did this threefold pattern come about?

TECHNOLOGY FOR ITS OWN SAKE:
THE PITFALLS OF BEING A WIZARD

Ironically, the strength of the Wizards – the often brilliant design-ers of high-tech products and systems today – is also partially responsible for their downfall: since they have so much scientif-ic and engineering expertise, they tend to think that everyone knows as much about technology as they do.[4] People who design things like playing with gadgets and figuring things out. It's a game to them and the more they do it the easier it gets for them. Some even like reading owner's manuals. And those with sophisticated technical skills are exceptionally good at discover-ing how complex devices work, which is why they got hired as designers in the first place.

But most of us aren't like that. We don't want to figure out what all of those buttons do, or why they are set up the way they are. We just want to get on with our lives or do our jobs well. When we make use of technology, we want to focus on achieving our goals, not on deciphering the technology. The design should be in the background of our attention. When we're dealing with a VCR, likely the main thing on our mind is recording a movie. We don't want to become computer pro-grammers to do this. The same thing holds true for more com-plex and potentially dangerous systems, like health care. Nurses choose their careers because they like to take care of people, not because they like to program complex computer-based medical devices or because they have a Ph.D. in computer science.

Yet the Wizards, who create products and complex techno-logical systems, frequently *are* computer programmers and sometimes *do* have Ph.D.s in the sciences or engineering, and it's very easy for them to forget how the rest of the world thinks. The result is often technological systems that are technically sound and easy for other designers to use, but that bury ordinary people in a quagmire of complexity.

Take the Infinia 7220, for example. This gadget was introduced by Toshiba with a great deal of fanfare in 1998.[5] It's the electronic equivalent of a Swiss Army knife: it has a TV, a computer, a telephone, a pager, a fax machine, a digital video disc player and a surround-sound movie and game center – all in one. It probably took an entire team of engineering geniuses to make it work. But if many people think operating a VCR is hard, can you imagine what it would be like to operate this technological octopus? The complexity would be way beyond the reach of the vast majority of people. No wonder Toshiba called the device Infinia – that's probably how long it takes to figure out how to use it.

Unfortunately, this Cyclopean Mechanistic trend toward bewildering complexity is intensifying. Take the lunatic example of the 2003 BMW 7 series, which has an electronic dashboard system called iDrive, that offers something like seven or eight *hundred* features. Even the company executives don't know the exact number, according to a report in *USA Today*.[6] Granted, a great deal of scientific and engineering knowledge was required to pull it off. But the BMW 7 Series is a car, not a spaceship. Is the end result something that most people can easily use? *Car and Driver* magazine called it "a lunatic attempt to replace intuitive controls with overwrought silicon, an electronic paper clip on a lease plan. One of our senior editors needed 10 minutes just to figure out how to start it."[7] An editor at *Road & Track* agreed: "It reminds me of software designers who become so familiar with the workings of their products that they forget actual customers at some point will have to learn how to use them. . . . Bottom line, this system forces the user to think way too much. A good system should do just the opposite."[8] As a result, *Road & Track* wound up entitling their review article: "iDrive? No, you drive, while I fiddle with the controller."

What on earth were the Wizards thinking when they designed this contraption? Simple: they were thinking about the gadget, not the driver.

And then there's the example of the London Ambulance Service – more significant since, lamentably, the Mechanistic tendency to focus its one-eyed gaze on gadgetry goes beyond everyday technologies to large-scale, safety-critical systems. In the wee hours of the morning on October 26, 1992, a new computer-based information system was introduced in London to help dispatchers assign ambulance crews.[9] The ambulance service is charged with responding to emergency phone calls from the 6.8 million people living in the 1,500 square kilometres in and around the city – an enormous public health responsibility. The new system was very ambitious. According to the later public inquiry report: "The concept behind the system design was to create, as far as possible, a totally automated system whereby the majority of the calls . . . would result in an automatic allocation proposal of the most suitable ambulance."[10] No system had ever before attempted to take computer automation of ambulance dispatching to such heights.

That first morning, when the number of calls was low, there were no noticeable problems. Everything seemed to be going according to plan. But as the call volume mounted, strains appeared; the computer algorithm wasn't doing a good job of assigning ambulances to calls. Within a very short time it became clear that all hell was breaking loose: multiple vehicles were being sent to the same incident, vehicles were being sent long distances when there were closer vehicles nearby, there were long waits, and callers started phoning back to the call centre, adding to the number of calls that the computer had to deal with. The dispatchers were in a panic, but their computer screens were swamped with messages showing how many calls were waiting to be serviced. They were unable to acknowledge every message because there were just too many, and soon there was a flood of new messages telling the dispatchers what they were already painfully aware of – that they weren't keeping up with the pace of events.

We can't blame the dispatchers. The information system had been designed from a one-eyed Mechanistic perspective to minimize human involvement, but it had the opposite effect. The Wizards hadn't thought of building in a way to identify duplicated messages – it never occurred to them it would be necessary. Incoming messages began to actually scroll off the top of the dispatchers' screens as the number of callbacks grew more insistent. The computer system became overloaded and slowed down. The ambulance response time increased. At the peak of the confusion, the response time reached a maximum of over three hours (the longest allowable response time was supposed to be seventeen minutes). The Wizards had not foreseen any of this. As the inquiry report said later: "the computer system itself did not fail in a technical sense . . . the system did what it has been designed to do."[11]

Some semblance of order was at last restored when people took over from the computer, but not before a very heavy price was paid. According to newspaper accounts, twenty to thirty people may have lost their lives as a result of the problems created by the introduction of the new information system.[12]

While knowing too much about technology for their own good, the Wizards also tend to know too little about the tasks that other people perform with technology.[13] Say designers are developing an electric guitar. Unless they're guitarists themselves, they're not going to know what's really required. What's difficult? What's easy? What's irrelevant? What's essential? They don't know. They can take a guess, but chances are they'll be wrong.

To pass on this lesson when I teach an introductory engineering design class, I use a deliberately contrived situation by asking students who are not guitarists to design an electric guitar together. The inevitable result of this class exercise is a user-hostile guitar. Because the students generating the designs know too little about playing a guitar, they have no choice but to make decisions on an ad hoc basis. One year, the class spent a long time arguing about how many knobs they should put on their electric guitar. "Two knobs," one student said. "No, three," volunteered another. "No way. An electric guitar has to have at least four knobs," said a third student. The argument went on and on. Finally, one student said, "What are the knobs for?" In class, I'm being deliberately simplistic to make a point. But it isn't far-fetched to extrapolate. If most engineering students have such a hard time anticipating user needs for a comparatively simple product like a guitar, imagine how much more difficult the job is for Wizards designing a safety-critical complex system.

HARD VS. SOFT

But how can we explain the second observation on our list – that the softer, non-physical aspects of technological systems can also make people's lives more difficult than necessary. In part, this is yet another natural outcome of putting the Wizards in charge of the design process. They're trained to focus on hardware and software, so the "softer" aspects of technology, such as work schedules or team coordination, simply fall away from the focus of their attention, if not their expertise.

The opposite approach can also backfire. The Humanists too can take matters to extremes. Instead of expecting too much of technology, they expect too much of human beings. If the systems are inadequate, they expect human effort and ingenuity to take up the slack. This attitude prevails in the health-care sector, with the ridiculously long work hours in hospitals being an excellent example: physicians are supposed to be physically resilient and mentally tough enough not to make mistakes even while working more than thirty hours in one shift, and 120 hours in one week. That's pushing the Cyclopean Humanistic world view to an extreme – let's call it super-Humanistic – since it expects people to be superhuman. This idealistic over-estimation of human capabilities ensures that the "softer" aspects of technological systems don't get the attention they deserve during the design process.

THE LOGIC OF HUMAN DESTINY:
TRANSITIONAL INSTABILITY AS A HARBINGER OF CHANGE

The third observation on my list is particularly perplexing: if we already know the reasons technology is out of control – the reasons I've outlined above – why are things only getting worse? Why don't we just address the causes and regain control of technology, especially since there's so much at stake for all of us? Answering this question requires us to take another historical detour. Although technological systems have never been as complex and as troubling as they are now, increasing complexity and social decay aren't new kids on the block in the history of humankind. On the contrary, they're perennials in cultural evolution, as Robert Wright argued so well in *Nonzero: The Logic of Human Destiny*.[14]

Over the past twenty millennia or so, humankind tried out innumerable technical innovations in an attempt to improve things. Bad ideas fell by the wayside because they didn't result in social progress. One technology that made it through this Darwinian selection mechanism was the rabbit net. Yes, the rabbit net. It made it easier for people to catch food and thereby increase their chances of survival and reproduction. However, it took a particular social structure to take full advantage of this technology: bands of hunter-gatherers. One person alone just couldn't hack it. Rather than try to catch a rabbit on their own (by running after it?), some innovative entrepreneurs created a new form of social coordination – and these coalitions resulted in a shared benefit – more food – for those who joined the band (we can only guess at what happened to the solitary-minded who decided to go solo). The technology caught on, and more and more people formed into bands, and the human population flourished.

But after a period of productive growth, rabbit nets started to lose their appeal. The population was now growing at a faster rate, and so the need for food jumped accordingly. Rabbit nets were no longer capable of supplying the demand and in due course the rabbit-net-based way of life started to break down. As people searched for a better way, there was probably a great deal of what Robert Wright refers to as *transitional instability* – turbulent and frustrating times, during which new technologies are developed, tested, found to be unsatisfactory and discarded. But a viable solution has to appear eventually, because people just won't put up with chaos forever; human ingenuity and the never-ending quest for a better quality of life will ultimately breed order and progress.

The viable new technology proved to be farming. It was more complex than rabbit nets, but it had greater potential to feed a much larger population. And as people figured out how to take advantage of it, a time of productive growth ensued. But the now old-fashioned social structure, made up of bands of hunter-gatherers who had been capable of controlling the comparatively simple rabbit-net technology, wasn't sophisticated enough to make the most of the promising new technology. As people gained more experience with farming, they probably began to realize the old ways just couldn't get the job done any more. The resulting clash spurred the search for a new social structure capable of harnessing the more complex technology. Various candidates were tried out, and abandoned; chaos and strife likely prevailed as the inadequacy of the old way of doing things led to social decay or stagnation. But there was too much at stake to give up, and eventually a solution was found: agricultural chiefdoms. These new, larger coalitions required greater collaboration but the effort was worth it, because the result was a net benefit for all who participated: having food is good.

The rabbit-net to farming story beautifully illustrates a developmental cycle consisting of four logical phases. Robert Wright refers to this as *the logic of human destiny*. First, the last 20,000 or so years of human history has been influenced by increasingly sophisticated technologies: rabbit nets gave way to farming, writing, money, printing and then computers. Few people would dispute this claim. Second, the same 20,000-year period has also been influenced by increasingly sophisticated social structures: bands of hunter-gatherers gave way to agricultural chiefdoms, then city-states, market economies, nation-states and the global village. This too appears straightforward.

The third phase of the four-part picture is where it gets a bit complicated. Robert Wright's claim is that technological innovation is chaotic, and only after a Darwinian survival-of-the-fittest process of testing and elimination do successful new technologies become established. The selection mechanism requires both of the first two processes to combine in a reciprocal self-organizing pattern. As Wright says: "Your brain may give birth to any technology, but other brains will decide whether the technology thrives. The number of possible technologies is infinite, and only a few pass this test of affinity with human nature."[15] The innovations that don't pass this test get thrown on the scrapheap. But to be successful, candidate technologies need the support of social structures that foster cooperation and coordination amongst individuals or institutions. And as the complexity of the technology increases, so does the sophistication of the coalition required to support and control it effectively. Even larger webs of social interdependence are needed to harness the power of the latest technology that has found an affinity with human nature.

The important fourth part of this developmental cycle is the transitional instability that results as new technologies and social structures arise and are overthrown. This fluid phase is a transitory no-man's land; the traditional way of thinking has lost its appeal and is leading to social chaos, but a new way of thinking that can lead to social progress has yet to appear on the horizon (sound familiar?). And just when you think things are at their worst and society is totally out of control, real advances are most likely to take place. As Wright puts it: "Turbulence and chaos often turn out to be harbingers of new forms of order."[16] In other words, sometimes things have to get really awful before they can get better, but the bad is what fuels the search for the good, and so, time and again, the reciprocal evolutionary processes linking technological and social structures result in a greater benefit for an increasing number of people.

And while the world is indeed filled with gross injustices and there is tremendous room for improvement – which is why I decided to write this book – it seems safe to say that more and more people are now living, on average, better and longer lives than in the past. The road to cultural progress is littered with detours; sometimes we seem to be stalled, or even going backwards, before we move forward again – slavery, genocide, human sacrifice, war and deadly plagues come to mind. But these periods of transitional instability play a role in sparking changes to improve the greater good (although often far more slowly than we would like).

Transpose Robert Wright's logic of human destiny to today's technology, and I believe it does a remarkably good job of explaining why our troubles are increasing. The Cyclopean Mechanistic and Humanistic world views worked well when technology was less complex and changed more slowly, but in modern times they're no longer adequate. Consider the situation in health care. Thanks to people's remarkable ability to adapt to extreme conditions, long work hours in hospitals may have been acceptable at one time, but given the enormous complexity of modern health-care systems and the constantly increasing pressures they have to cope with, doctors and nurses are now stretched well beyond their breaking point. Expecting them to be saviours in today's challenging health-care climate, regardless of how systems are designed, is sheer folly.

Back when technology was comparatively primitive, the products or systems created by Mechanistic designers may have challenged human capabilities but rarely exceeded them. But now that the Wizards have infinitely more sophisticated and varied technologies to play with, their professional sandbox is much larger – naturally, if a designer is faced with many options, the temptation is to use as many of them as possible, particularly when the marketing department wants to use new features as a selling point, as they frequently do. And the pace of technological change also has a bearing on the viability of the old Mechanistic approach. When technology changed relatively slowly, people had time to adapt to products or systems that weren't designed with humans in mind. For example, the layout of the keys on mechanical typewriters was originally designed to slow typists down, because the typewriter parts would jam if the keys were struck too quickly in succession – the keyboard was a deliberate awkward fit between people and technology. And the fact that the layout was kept constant over the years (it still hasn't changed much) gave people time to adapt to the design and cope with it. But now that we're surrounded by computer technology that changes every couple of years, we can't keep up; most of us are perennially lagging behind. In a sense, we're stuck in the intellectual straitjacket imposed by antiquated social structures that were once useful, but are just inadequate for confronting the increasingly complex and dynamic technological systems that power our modern world. Just as bands of hunter-gatherers couldn't deal with farming, the Cyclopean Humanists and Mechanists can't cope with so much change and complexity.

The result is just what Wright would predict: we're experiencing some big-time, nasty, transitional instability in the technological world – technology is wreaking havoc all around us. But until a new and better way of thinking crystallizes and takes hold, we'll keep on resorting to familiar but outdated ideas, because they used to work and they're all we have in our conceptual tool box. And given the lessons of history, things will have to get *really* bad before we let go of those old ways of thinking. All the signs tell me we've now reached that point of intolerable but fertile transitional instability.

We need a new approach, one that's better tailored to modern times, more sophisticated than the old Mechanistic and Humanistic world views, to deal with the new technologies of the twenty-first century. But while focusing on technical details alone, or on people alone, won't cut it any more, we don't want to throw the baby out with the bathwater, as my grandmother would have said; getting the electrons going in the right direction is just as critical as it's always been, and people do have an important role to play in technological systems.

A COMMON SENSE SOLUTION: THE HUMAN-TECH REVOLUTION

So where shall we look to find an alternative, new world view that's up to the task? The poor design situations I've described offer a clue. They all have one thing in common: they make unrealistic assumptions about human beings, creating a *bad fit* between people and technology. This statement is true whether we're dealing with physical objects resulting from a Mechanistic perspective, such as the BMW iDrive dashboard, or with non-physical entities resulting from the Humanistic perspective, such as medical work schedules.

There's something very odd going on here. If designers made completely unrealistic assumptions about the physical world when designing technology, then we would blame them (and would likely sue them) for technical incompetence. Yet when they make grossly unrealistic assumptions about human nature – expecting us to make sense of eight hundred commands while driving a car, or take other people's lives in our hands and never make a mistake while working 120 of the 168 hours in a week – we don't blame the designers, we blame the unfortunate people who are just trying to do what the design requires. This makes no sense whatsoever. And it needs to change.

The design of technology, as Wright indicated, has to have affinity with human nature to be successful. I would further argue that design should begin by identifying a human or societal need – a problem worth solving – and then fulfill that need by tailoring the technology to the specific, relevant human factors.

In other words, design needs to include not just knowledge of the physical world, but knowledge of human beings as well. It may seem obvious, or good common sense, but it isn't happening very often. Just as the technical sciences have acquired a great deal of knowledge about the physics and chemistry of the world, the human sciences have also accumulated an impressive body of knowledge about the kinesiology and psychology of human beings. People are more adaptive, and thus more forgiving, than particles, but there are still many things that we *know* people can't do very well, or at all, and those things should be avoided by designers. So if we can tailor technology to our physical world, there's no reason why we can't also tailor it to human nature. Doing so should lead to a seamless integration of people and technology, eliminating the bad fits that are causing so much trouble. And of course, we should ensure that the design of the technological system is *problem-driven*, that it aims to fulfill a human or societal need, so that we avoid the Mechanistic tendency to design technology for its own sake. Rather than thinking about the Cyclopean abstractions of "technology without people" or "people without technology," we can focus our attention on what matters most – the people-technology *relationship* as it affects human and societal needs.

Some readers will recognize what I'm advocating as an example of *systems thinking* – a holistic, problem-driven way of looking at the world, an approach that focuses on *relationships* between system elements, whatever form those elements happen to take (in our case, people and technology). Rather than following the old Laplacian reductionist doctrine of carving things up into smaller and smaller pieces and examining each tiny element in detail and in isolation – the kind of thinking that got us into this mess in the first place – systems thinking focuses on the big picture, the interactions between the elements. This way of tackling problems is slowly emerging as a useful approach in many areas of society, so the vision I'm proposing here is consistent with revolutions that are going on elsewhere. But systems thinking is still a minority view and many people have never heard of it, so I'll take a few moments to describe it.

Think of a system as a whole consisting of interacting parts – something that "talks" to itself. The parts can be anything: a husband and a wife; the components in a car; a predator and its prey; all the animate and inanimate objects in our biosphere; the buyers and sellers in a financial market; or the set of countries belonging to the United Nations. It's the relationships that are the key. And therein lies the slippery nature of systems thinking: a relationship isn't a physical object that you can hold in your hand; it's an *emergent property*, a gestalt, which only comes into existence when the parts it comprises are brought together and configured in a particular way.

Here's a very simple experiment you (and your children) can try at home to better understand what I'm talking about. Get someone to build an apple-and-orange system by holding an apple and an orange right in front of you. OK, are you ready? First, grab the orange. No problem. Now, grab the apple. Again, easy. So now try to grab the distance between the apple and orange. Having trouble? Try again. No matter how quick your reflexes, you can't do it – there's nothing there to get hold of. The distance is real, of course, but it's intangible; it doesn't have a material foundation.

This simple example is static, whereas most of the relationships that govern modern times are always changing, which makes things more complicated. And this emergent property – the distance – doesn't exert any influence on the behaviour of the system, whereas in the more sophisticated systems that we'll be looking at, relationships act like an invisible force propelling people's behaviour (for good or ill). We see this in everyday life: as I'm writing these words, the radio is telling me that people in Cambridge, Massachusetts (where I'm currently living), are flocking to the stores to buy winter boots, rent videos and acquire shovels. Why? Because we've just been told that later today and tomorrow we're going to be hit by the biggest blizzard of the season, burying us in sixty centimetres of snow. This relationship between the weather, the needs it creates and people's behaviour is intangible – I can't see it or touch it – but it's real and it's powerful. Just ask the salespeople who are frenetically trying to keep up with the mad dash of customers to their stores today.

So now you know what an emergent property is – not an object per se, but a relationship between system elements. And this is a priceless lesson at a time when we live in a dynamic relationship with almost everything around us. The world of people and technology is a vast ocean filled with such relationships, and one of the most important aims of this book is to help you see these relationships – good fits, bad fits – that exist all around you every single day of your life, rather than pass them by without even realizing that they're there, let alone that they're affecting your quality of life.

We can use a variation of my simple experiment to reinforce the limitations of reductionism. First, ignore the orange completely and look just at the apple in isolation. Now, try to find the distance between it and the orange (don't cheat by sneaking a peek at the apple). You'll never find it. You can even go to the kitchen, grab a knife, carve the apple into tiny pieces, put those pieces under an electron microscope or try to decipher the apple genome, and you'll never, ever find that slippery distance between the apple and the orange. Why? Because it's a relationship – an emergent property – that falls outside of the conceptual blinders imposed by the reductionist approach I forced you to take. That's why reductionism can make us half-blind.

Why does systems thinking get around these limitations? The answer is so simple that a five-year-old can understand it. As James Gibson – a famous American psychologist who advocated a systems approach to his own discipline[17] – purportedly said: "If you don't take it apart in the first place, then you don't have to put it back together again." In other words, if we didn't carve up the apple-orange system and study the apple in isolation, we wouldn't be faced with the problem of building up knowledge of that elusive distance relationship from knowledge of the apple alone. In the context of the problems in this book, this means that if we didn't take apart human-technology systems in the first place and study technology in isolation or people in isolation (as the Cyclopean Mechanistic and Humanistic world views encourage us to do), we wouldn't have to struggle to understand why there are so many bad fits between people and technology, and we wouldn't be faced with the challenge of replacing those bad fits with good ones – we wouldn't have to put Humpty Dumpty back together again.

The systems approach encourages us to think about relation-ships between people and technology. Design is all about creating emergent properties. That might sound weird, but it's true. Different system outcomes can be had by building different relationships, sometimes using the very same or similar parts, because small changes to the parts can make a tremendous difference (I'll present many examples later). So what we need to do is design relationships with technology that lead to har-mony, not tension – good fits, not bad. And we need to base those new designs on a broad, powerful and proven set of systems-thinking techniques.

I wish I could take credit for inventing systems thinking, but I can't. Adam Smith, the Scottish economist and author of *An Inquiry into the Nature and Causes of the Wealth of Nations,* unknowingly became the first systems thinker in 1776 (the term didn't exist back then) when he explained that the behaviour of markets can only be understood by looking at the *relationships* between buyers and sellers, not by homing in on the buyers and sellers in isolation. In fact, his famous concept of the "invisible hand" is a prototypical example of an emergent property – you can't touch or hold it, but it exerts an enormous influence on the overall behaviour of the system nevertheless (as financial analysts can attest). This concept was a landmark in intellectual history, and its implications are still being mined. What I've tried to do here is take systems thinking and apply it to our troubles with technology at all levels of human activity. In the remainder of this book, we'll encounter many modern-day technological equivalents of Smith's invisible hand, some that cruelly push us over the edge of safety to disaster, and others that gently lift us to awesome heights, from which we can measure the achievements that humankind is capable of, given a benevolent set of conditions.

Because a systems approach to the design of technology is a fundamentally different way of thinking, a departure from the well-trodden Mechanistic and Humanistic paths, there isn't a catchy, familiar term to describe it. So I had to invent one: *Human-tech*.

HUMAN-TECH

You could say that the term itself was created using a Human-tech approach; it's easy to remember because it consciously mimics the conceptual structure of the idea it refers to – what you see is what you get. First, it's a compound word made up of two parts to remind us that people and technology are both important aspects of the system; we want to have binocular vision, not remain half-blind. Second, the hyphen connecting the two parts highlights the importance of the relationships between humans and technology – it's a visible stand-in for those otherwise invisible (but crucial and powerful) emergent properties. Third, "Human" comes first to remind us that we should start by identifying our human and societal needs, not by glorifying some fancy widget in isolation; "tech" is a means, not an end in itself, so it comes second. Fourth, "Human" is capitalized and thus more salient to remind us that designs should have affinity for human nature, something frequently over-looked by the Wizards; "tech" is in lower case and thus less salient because everybody already knows that the technical details have to be tailored to the laws of physics and such.

And finally, "Human-tech" provides an essential addition to two other terms we all know well: *low-tech* and *high-tech*. In the old-fashioned Mechanistic way of thinking, low-tech is usually seen as primitive and therefore bad, and high-tech as advanced and therefore good. From a Human-tech perspective, this isn't necessarily so. Low-tech (such as a pencil) could also be Human-tech, because it's easy to write with and doesn't require periodic rebooting; conversely, high-tech (such as the BMW iDrive) may not be Human-tech, because it can be awfully hard to find the right command out of eight hundred using a sequential, hierarchical menu command structure while simultaneously negotiating bustling traffic in a big city (or, Lord forbid, when you're about to enter the Magic Roundabout – of which more later). From our new perspective, the old distinction between low-tech and high-tech is completely irrelevant – except in so far as it's pertinent to our human needs and capabilities. High-tech can *sometimes* make it easier to create Human-tech, by providing new and useful possibilities – the Internet has done wonders for society, as I'll discuss later – but just as frequently it does the opposite, complicating every move we make and threatening our quality of life.

So there you have it: a new, reader-friendly term for a new way of thinking that owes a nod to Adam Smith's genius and has parallels with the systems thinking revolutions that are slowly transforming other areas of society. My deepest professional hope is that this simple word – Human-tech – will help to power a conceptual revolution, and tear down the roadblocks put in our way by antiquated Mechanistic and Humanistic ways of thinking. A Human-tech Revolution would completely change how we live with technology, and would do away with the transitional instability that currently engulfs us.

I've always been an optimist.

Some readers may feel that Human-tech thinking isn't revolutionary at all, but just plain old common sense. They're right (sort of). As Arthur Koestler suggested in his book *The Sleepwalkers,* psychological roadblocks can obstruct our vision of reality and make it excruciatingly difficult – sometimes impossible – to see the obvious. The division of knowledge between the one-eyed Mechanistic and Humanistic world views provides a magnificent example of what Koestler was talking about. If the Human-tech view had become the common-sense approach, then I wouldn't have had to invent a new word for the idea – and I'd be hard pressed to find any examples of bad fits between people and technology. Only when we look at the world without the blinders imposed by outmoded views does the importance of designing technological systems to have affinity with human nature appear so obvious.

A few sectors of our modern world – I'll discuss them as models to follow – have already experienced a Human-tech Revolution, and have used this new world view to create successful technological systems that satisfy rather than exacerbate human and societal needs. However, most of our world is still in the dark ages: we're still putting our blind faith either in people alone, or more typically in technology alone. Human-tech thinking is still a well-kept secret, but the cat needs to be let out of the bag, because there's so much at stake for each and every one of us. A new vision of society – one that bridges the technical and human sciences – has the potential to improve the quality of life for generations to come, but only if we act now. If we fail to act, irreparable harm may be done to human life and our world.

A CONCEPTUAL ROAD MAP: THE HUMAN-TECH LADDER

If we're going to follow the Human-tech approach, we have to develop a good understanding of the principles that govern human behaviour. Otherwise we won't know what shape our design solution should take, and the ideal – affinity with human nature – is unlikely to ever be attained. The apparent simplicity of this concept belies its difficulty – humans are *very* complicated beings.

There's more than one way to get a handle on this complexity, but the first and best approach is to adopt a problem-driven focus – a Human-tech focus. If we consider only the human characteristics that are relevant to the specific design problem we want to solve, everything becomes much, much simpler. For instance, if we're designing a toothbrush, then only a very thin slice of human nature is relevant to our concerns. We don't have to worry about people's political values, for instance, since they're unlikely to affect the design problem at hand. Adopting a context-specific, problem-driven approach narrows down the amount and kind of knowledge about human behaviour that we need to consider to find an effective design solution. In

	THE HUMAN FACTOR
	Political
	Organizational
	Team
	Psychological
	Physical

Table 1. The human factor can be categorized into five levels.

other words, the Human-tech approach doesn't require an encyclopedia that collates every single known fact about humans; we can get away with a more practical designer's handbook that focuses solely on the human aspects that are most important for the particular technological system under consideration – a much smaller number of facts.

But even if we narrow our focus in this way, leveraging knowledge of human behaviour can still be a daunting task. A good way to make the Human-tech approach more tractable and get a grip on this complexity is to organize our knowledge of people systematically, in a multi-faceted way. We know a great deal about people from a diverse set of perspectives, including the physical, psychological, team, organizational and political. Table 1 shows how we can organize this knowledge into five levels, each representing a relatively distinct aspect of people, and thus, a qualitatively different way of thinking about them.[18]

At the lowest level, we can view people strictly in *physical* terms. Individuals differ, of course, in shape and size of their bodies, their physiology, strength and dexterity, but for any particular design context, certain capabilities and limitations will likely be shared by the vast majority of the intended users of that design. Since the objective of the Human-tech approach is to design *to* human nature – to build a harmonious relationship – the solution we develop must respect these physical characteristics. If we're creating a product for children, the design shouldn't require them to reach two metres above the ground, because children can't reach that high. If we're creating a manufacturing system for workers, then the design shouldn't require them to lift three thousand kilograms manually, because no human being has the strength to lift such a heavy load unaided.

Not only does the physical level tell us something about human capabilities, but it can even inform us about specific recurring problems that have already been identified and perhaps even how to prevent them from cropping up again. Repetitive strain injury (RSI) is an example. There's a great deal of useful knowledge about how RSI is caused, and about how the risk of RSI can be reduced if we adapt the design of workplaces to conform to people's physical capabilities.[19] If the human factor is taken into account, a tight fit between person and design can be achieved and the technology is more likely to fulfill its intended purpose.

Even if a design is a good fit in physical terms, there may be mismatches at other levels. For instance, depending on the product or system to be designed, we may also need to think about the *psychological* perspective. Some of the potentially relevant principles here include the abilities of our short- and long-term memory, our intuitive expectations for making sense of the world, our (limited) ability to perform complex mental calculations, and our exquisite capacities for pattern recognition. Again, a tremendous amount of variability can be observed across individuals, but for a specific target audience there will be basic psychological characteristics that can be used as a litmus test for any Human-tech design. If we're designing a steering wheel for a car, then turning the wheel to the left shouldn't cause the car to move to the right because that would run counter to the normal mental expectations about cause-effect relationships. If we're creating software for a computer, the design shouldn't require people to memorize twenty new pieces of information simultaneously while in the middle of performing a task, because we know that human short-term memory can only hold about seven pieces of information at once. A design that ignores these fundamental psychological characteristics will fail. Cars with backward steering wheels will be driven into ditches and computers that require mental gymnastics will lead to mistakes that cause us to pull our hair out.

There are many well-known psychological limitations that a Human-tech design can be tailored to overcome. One such limitation is the natural human tendency to seek out (and thus find) information that is consistent with our views and to ignore (and thus fail to find) information that could falsify our views – what is aptly called "confirmation bias."[20] Designers who take the trouble to identify these psychological foibles can design against these dysfunctional patterns of mental behaviour.

It's important to note that the psychological level is qualitatively different from the physical. A design can provide a great fit in physical terms yet fail miserably in psychological terms, and vice versa. The difficulties we encounter with many computer software programs are rarely physical. Almost all of us can reach the keys on the keyboard, to press down on those keys, and read the characters on the screen. Unless you spend too much time on your computer and get RSI, there's usually no problem at all with these physical factors. Rather, our difficulties usually lie in the psychological realm. For example, we may not be able to recall the label for a particular command, or we may not be able to find the command we're looking for in the myriad options provided under the many counterintuitive menus.

Why, for example, do we have to select the *Start* menu to find the command to *shut down* the computer?

The physical and psychological levels do a relatively good job of describing an individual person, but people don't usually work in isolation. Many jobs require us to function in a *team* (or group) consisting of two or more people who have to communicate with each other and coordinate their respective actions to achieve individual and common goals: airplane pilots, nuclear power plant operators, health-care providers and many other professionals, including office workers, all work in teams. Team dynamics introduce a host of new issues that a Human-tech design must take into account on top of physical and psychological constraints. For instance, if teams are to function efficiently, explicit goals and priorities should be identified and communicated among the members so that they can all paddle in the same direction and thereby achieve a synergistic effect, as opposed to having each member go off in a different and conflicting direction and consequently dilute the team's effectiveness. To take an absurd example, it would be a bad idea to have one cockpit at the front of an airplane for the captain and another at the back for the first officer with no video or audio hookup between the two. Even if the design of the two cockpits was physically and psychologically appropriate for the two pilots, it would be a disaster for the team, because it would be virtually impossible to achieve effective coordination; one pilot might be trying to land the plane while the other was trying to hold it at cruising altitude. Indeed, research has shown that if team members can see the actions of their colleagues with minimal effort, sometimes just a quick glance out of the corner of an eye, they will be able to coordinate their respective actions even while they're busy with their separate tasks.[21] Many very expensive and technically sophisticated video-conferencing systems have failed precisely because they didn't respect this fact about people at the team level.

And as with the previous two levels, the team level also carries with it certain recognized weaknesses, such as groupthink.[22] Teams of individuals have been known to converge on a particular interpretation of events just to achieve team harmony and then become fixated on that interpretation, ignoring other possible explanations. If designers are aware of these tendencies, they can try to minimize them.

But, of course, teams don't usually exist in a vacuum. They're usually part of a larger unit, an organization composed of many teams or other types of organizational units. This *organizational* level introduces yet another set of human characteristics. Vision and leadership, incentives and disincentives, and the way information flows are among the factors that have an impact on organizational behaviour and provide a fourth level for establishing affinity with people. For instance, if an organization wants to learn from experience, it should avoid creating a reward structure that punishes and blames people for bringing forward bad news. "Shooting the messenger" will only encourage people to cover up problems, thereby suppressing information that could be productively used to improve performance. This is a basic principle of human behaviour at the organizational level.

Another well-known propensity that can lead to a negative outcome is the sunk-cost fallacy – the tendency to invest more resources to try to recuperate resources that have already been lost.[23] Instead of cutting their losses after a bad initial decision, organizations sometimes feel that they "can't back out now" and continue to waste resources by investing in a losing cause. The sunk-cost fallacy can also affect individuals and governments; just think about persistent gamblers or the United States' delayed decision to pull out of Vietnam. By knowing such pitfalls exist, system designers can try to create organizational safeguards to minimize their likelihood.

System design decisions at the organizational level can also impinge on lower levels. The problem of medical fatigue I described earlier illustrates this. Decisions about staffing and shift-work schedules are usually made at the organizational level, but they can impact the psychological or physical levels. If the number of nurses assigned to a hospital ward is too low, the work load of an individual nurse may push his psychological ability to cope to the breaking point. By overtaxing their workers, health-care organizations are increasing the likelihood of errors that can injure or kill patients. Similarly, if medical residents are assigned to work very long shifts, then they may become drowsy and inattentive, a physical state that's unlikely to advance the overarching aim of patient safety.

The topmost level in our framework is the *political*. Here, there are basic considerations, such as public opinion, social values and cultural norms, that must be respected. Any legal or governmental design that does not pass the test of affinity with human nature will be unlikely to achieve its ends. For instance, think about the lessons learned during the Prohibition era of the 1920s and 1930s in the United States. A law that prohibited the sale or consumption of alcohol was passed in an individual-istic culture that valued freedom and had always accepted drink-ing as a social activity. Naturally, the law failed. Making drinking illegal didn't stop it at all, and instead led to all kinds of dys-functional side-effects, like bootlegging and a sizable new source of income for organized crime. Designing Prohibition legislation in such an environment is the political equivalent of designing a manufacturing system that requires workers to lift three thou-sand kilograms with their bare hands; the intended purpose won't be achieved because there is a fundamental mismatch between the design and the human factors. If the one case is physically impossible, the other is politically impossible.

Granted, our understanding of human nature at the political level isn't as sophisticated or as complete as it is at the physical or psychological levels, because the phenomena to be explained are far more complex. Nevertheless, there are well-known patterns of dysfunctional behaviour that can be used as bases for system design, such as "the tragedy of the commons," which states roughly that if all individuals are given unchecked freedom to pursue their self-interest in a shared, resource-limited environment, those resources eventually will be depleted to everyone's detriment.[24] Politically imposed restraint is required to avoid the tragedy. Knowledge of this type can be used by designers to develop legislative controls, for instance, that keep society from running amok.

This observation underscores a point I made earlier – design shouldn't be concerned just with the creation of hardware and software. At the lower levels of our framework, design is concerned with creating physical or virtual objects that are a good fit with human physical and psychological characteristics, such as a door handle that's placed at hand height and is easy to grasp, or computer software that's easy to understand and operate. These are the kinds of examples that normally come to mind when we think about designing technology. But as we move to higher levels, the focus of design is no longer just "stuff" or "widgets." To match what we know about human behaviour at the team, organizational and political levels, we also have to think about things that exist as patterns or concepts, such as authority relationships, staffing policies, laws and regulations. Most people don't think of these abstract "objects" as having anything at all to do with technology. But although they're not tangible, they are real, and they have important consequences for the successful functioning of technological systems.

Political objects like budgets, laws and regulations should be viewed as part of the technological system because they affect how well the physical objects in the system will function. Advanced medical devices would be dangerous if there were no laws governing the qualifications of the doctors who use them. Even if such a device were technically perfect and had been carefully designed to have an affinity with human nature in physical terms, in the hands of a quack it could still physically hurt or kill people. The solution to this problem would not be to redesign the machine, but to design more stringent professional regulations. Indeed, as we'll see later, the political level is perhaps the most important of all because it influences, for good or ill, all the other levels. That's why complex technological systems are frequently referred to as sociotechnical systems; they're made up of both technical and socio-political elements.[25] As a result, for large-scale complex systems, "design" must be interpreted in a very broad sense, taking in both the technical and the administrative aspects of the overall system. In either case, the designer's key objective is to dovetail all parts of the technological system with the behaviours and abilities of the human beings who will be operating it or affected by it, so that the overarching need that's driving the design can be fulfilled.

I've summarized the Human-tech relationships in Table 2. In the right-hand column are five key human factors and in the left-hand column is a partial listing of corresponding hard and soft technologies. The rungs connecting the two sides of the ladder represent the relationships we need to build into the system to achieve good fits. We start off by identifying a human or societal need (the overarching requirement); from there, we design technology that's tailored to our knowledge of the human factors that govern our behaviour. If we're trying to design computer software that's easy to use, the psychological factors might be the biggest consideration. If we're designing a very large, complex system – like a nuclear power plant or a national health-care system – we may have to take into account all factors, at each level.

The Human-tech ladder will serve as a conceptual road map for the journey we'll be taking in the remainder of the book. We'll begin at the bottom of the ladder to look at how technology either works for people or wreaks havoc at the basic level of physical demands and needs; and then we'll move up the ladder,

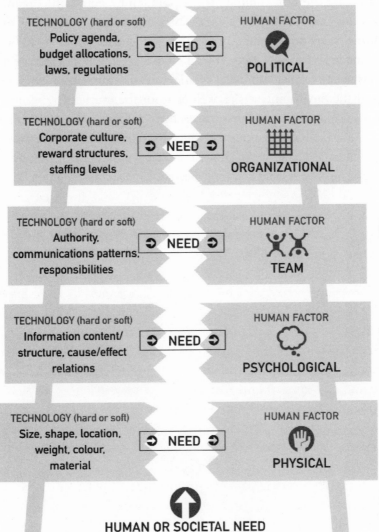

TECHNOLOGY (hard or soft)
Policy agenda,
budget allocations,
laws, regulations

⮑ NEED ⮐

HUMAN FACTOR

POLITICAL

TECHNOLOGY (hard or soft)
Corporate culture,
reward structures,
staffing levels

⮑ NEED ⮐

HUMAN FACTOR

ORGANIZATIONAL

TECHNOLOGY (hard or soft)
Authority,
communications patterns,
responsibilities

⮑ NEED ⮐

HUMAN FACTOR

TEAM

TECHNOLOGY (hard or soft)
Information content/
structure, cause/effect
relations

⮑ NEED ⮐

HUMAN FACTOR

PSYCHOLOGICAL

TECHNOLOGY (hard or soft)
Size, shape, location,
weight, colour,
material

⮑ NEED ⮐

HUMAN FACTOR

PHYSICAL

HUMAN OR SOCIETAL NEED
e.g.: the music revolution, the knowledge economy,
transportation, counter-terrorism,
public health, environment . . .

THE HUMAN-TECH LADDER

Table 2. The Human-tech ladder: Design should begin by understanding a human or societal need—and then tailoring the technology to reflect specific human factors.

applying the same Human-tech approach to teamwork, organizations and political systems. Throughout, I'll be drawing on some fascinating examples – both good and bad – that I've come across in the safety-critical sectors I described in the previous chapter: nuclear power, aviation, health care, airport security and the environment – as well as from the everyday consumer products that help make our lives heaven or hell. In every case, I will focus on the relationship – successful or otherwise – between people and technology. The Human-tech world view is all about designing for people and their needs.

Part Two

TECHNOLOGY FOR HUMANS

3

Let's Get Physical: Fitting the Design to the Body

THE HUMAN-TECH LADDER

TECHNOLOGY (hard or soft)
Size, shape, location,
weight, colour,
material

⊃ NEED ⊃

HUMAN FACTOR

PHYSICAL

At the basic, physical level – the lowest on our ladder – we may seem to be asking too much of a holistic, relational Human-tech approach. After all, the only leverage points we have at this level are elementary design attributes such as size, shape, location, weight, colour and material. How much impact can we really have on our quality of life by making such simple changes to a design?

The human body is multi-faceted, with many diverse characteristics, some small (the size of our hands), others big (the overall shape of our bodies), some relatively static (how tall we stand), others more fluid (how far we can reach), some anatomical (the number and location of our teeth), and others physiological (our manual dexterity and our basic bodily functions). There is a cornucopia of characteristics to consider.

Most of us don't give much thought to the diverse and ample package of physical capabilities that we carry around with us on a daily basis. Why should we? It's only when we can't successfully perform an everyday task – reaching for a distant object, lifting a heavy load, grasping an awkwardly shaped tool – that we stop and think about our physical capacities. And even then, we tend to look at ourselves in a negative light – I wish I had longer arms, I wish I were stronger, I wish I were more dexterous. Because we tend to look inward, reflecting on our bodily selves rather than outward at the design of the technology around us, we don't realize that our apparent shortcomings aren't necessarily signs of human frailty – the proverbial attribution of our failures to "human error." In many cases, the problem is that technology hasn't been designed to fit our bodies, even though the knowledge to do so has been available for some time.

THE HUMAN HAND:
A USER-HOSTILE TOILET-PAPER DISPENSER

Consider the human hand (a real one, not Adam Smith's invisible one). It comes in different shapes and sizes, but with rare exceptions it has certain invariant properties: at its core, it's made of bones and tendons, the bones are surrounded by pliant but vulnerable flesh, the flesh has an outer coating of smooth but also vulnerable skin, there are five fingers of different but fixed lengths, and each of those fingers is in a predictable position relative to the others. The fingers bend well in some directions, not so well or not at all in other directions, and are connected to a palm, which is connected to a wrist, and the whole ensemble has limited strength. This is not rocket science – a five-year-old would probably come up with much the same thing if you asked her to describe her hand. And if you want a more precise account, there are books where you can look up the typical range of dimensions for hands (and other parts of the human body).[1] There is even a science – ergonomics – with textbooks devoted to documenting the size and shape of human hands, along with the other physical properties of people.[2] Make no mistake – there is more information available about human hands than you will ever need or care to know about.

Given the availability of all this knowledge, it would make sense for product designers to sculpt the physical dimensions of products to match the known physical capacities of our hands – to build a harmonious relationship between the two. After all, it's comparatively easy for designers to change the dimensions of a product on a blueprint, especially since computer-aided design software has replaced manual drafting. In contrast, unless we have friends who are plastic surgeons, it would be pretty difficult for most of us to change the size, shape, reach or dexterity of our hands, even in the unlikely event that we wanted to. Adapting the design to the hand makes infinitely more sense than adapting the hand to the design.

But do designers actually do this? Well, not always. In one of my undergraduate classes, I ask students to look around their homes and come up with examples of designs that are difficult or frustrating to use. They've come up with some truly memorable examples, including one where the human hand was the main protagonist and a nasty toilet-paper dispenser was the main antagonist. One student reported on a particular dispenser that stores the roll of toilet paper in a metal box. In the opening at the bottom of the box where the paper comes out, there's a jagged, sawtoothed edge to tear it off. However, as the student pointed out, sometimes the paper isn't sticking out of the box as it should, and you have to reach inside the box to find it. Since the opening is at the bottom of the metal box and the gap is narrow (to discourage toilet-paper thieves?), you have to bend your hand and fingers to reach inside. This isn't something you look forward to doing when you're desperately in need of paper. The roll is easy enough to find by touch, but sometimes you have to rotate it to locate the edge of the paper. As you feel your way around inside the box in search of the paper, that jagged, saw-toothed edge is digging away at your fingers or the palm of your hand. We shouldn't have to contort our hands and risk getting cut up when using a toilet-paper dispenser. There has to be a better way to get the job done (so to speak).

Mind you, if we pretend for a minute that we don't know anything at all about human hands and focus just on the technical details of this design, we would conclude that there's not much wrong with the toilet-paper dispenser. It's mechanically sturdy, having been crafted to easily withstand the weight of a roll of toilet paper. Also, the dispenser's dimensions have been carefully designed so that a brand new roll of toilet paper fits snugly inside the box. Moreover, the jagged, sawtoothed edge isn't at fault; it's been deliberately created to be sharp enough to tear the paper. The elegance and forethought put into the purely technical aspects of this design lead to an obvious question: if designers can fit the design of the dispenser to the physical characteristics of a roll of toilet paper, why couldn't they also fit the design to the physical characteristics of the human hand?

They could have. But they didn't, because the idea isn't obvious from a Mechanistic perspective, which puts strictly technical minutiae at the centre of the universe, and virtually leaves out the human factor. The psychological blinders imposed by this technocentric world view are so powerful that they effectively prevent adult designers from seeing what even a five-year-old could have told them about a human hand.

THE HUMAN BODY:
A MECHANICAL LATHE FOR MIDGET EXTRATERRESTRIALS

A generous sceptic could argue that hands are a small part of the human body: surely in situations where a larger, more salient part of the body is at play, designers would take the human factor into account. We could call this the "salient body-part hypothesis" – if it's big enough, it won't get ignored; if it's too small, it will.

Mechanical lathes that are used for machining metal parts provide a way of testing the "salient body-part hypothesis" because they require people to use their entire body. The machinists have to stand while operating the lathe. They have to use their torso and arms to bend in various directions to reach the lathe controls, and they have to use their hands to grasp those controls. With the whole body in play, the salient body-part hypothesis would dictate that human physical characteristics couldn't possibly be ignored in the design of the lathe.

In 1981 when I was an undergraduate engineering student at the University of Toronto, one of my very first (and very best) professors, Patrick Foley, tested the hypothesis for us. He described the design of a commercially available lathe. As with the toilet-paper dispenser, the designers had made sure that the strictly technical details were well taken care of; the lathe was perfectly capable of machining metal. But was the design tailored to fit the human body?

To answer this question, Foley put up a slide of a cross-sectional view of the lathe, showing where the controls were placed, and how far apart they were. To see if this layout was consistent with human abilities and limitations, Foley reverse-engineered the physical aspects of the human from the lathe design. By identifying what steps the machinist has to perform to do the job, and then examining the locations of the existing controls on that lathe, it was possible to determine who the lathe was (implicitly) designed for. What body dimensions would that person have to have to operate the lathe easily? Foley put up another slide, which vividly illustrated the answer. This one too showed the relative position of the lathe controls, but it also superimposed a drawing of the human that the design ideally called for. We would expect that the dimensions of that ideal human would be representative of a typical user, but they weren't even close.

That ideal user would have to be four feet, three inches tall, have shoulders two feet wide, and an arm span of eight feet. The figure shown in the slide was straight out of a science fiction movie: short, with ridiculously wide shoulders and arms that hung almost to his ankles. I still remember looking at that slide and imagining how difficult it must have been for the machinists to do their jobs, given the unrealistic assumptions the designers had made about human anatomy – the relationships built into the design led to glaring tension. Yet the (mechanically competent) lathe was commercially available and used by machinists, so somehow they coped, probably by contorting their bodies and walking back and forth to conform to the inflexible technology. So much for the salient-body-part hypothesis.

These two examples deal only with human needs that are a part of our work life (the lathe) or our daily personal life (the toilet paper dispenser). The damage caused by the Mechanistic perspective in these everyday situations is comparatively modest; nobody has ever been killed by a user-hostile toilet-paper dispenser or a lathe for short extraterrestrials. Can the same idea of creating a design to fit the human body in order to make life easier also be valuable for people who make life-and-death decisions on the job? Aviation provides a perfect testing ground, for if there's one place where split-second decisions can have a life-and-death impact, it's an airplane cockpit.

AVIATION SAFETY: AN "INVISIBLE HAND" IN THE COCKPIT

Flying used to be incredibly dangerous.[3] The first powered flight occurred in 1903, the first fatality in 1908, and the first midair collision in 1910. At that time, there were about two thousand airplane pilots in the world and 32 had already been killed. The U.S. Air Mail, founded in 1918, provides more evidence of just how perilous air travel was in those early days. Thirty-one of the first forty pilots were killed in service. When over 75 per cent of the people die from the technology they're trying to control, you know you've got a shockingly bad fit on your hands. Unfortunately, deadly aviation accidents continued to occur on a regular and all too frequent basis, not just at the dawn of the industry, but for several more decades as well.

Today, however, aviation is remarkably safe and getting even safer. Given the unforgettable mass murders of September 11th, you might expect that 2001 would have been a particularly horrible year for aviation safety. In fact, the accident data indicate otherwise.[4] There have been so many improvements to aviation safety over the years that despite the four horrible crashes caused by terrorist attacks in the U.S., the total number of major airline crashes around the world in 2001 fell to the lowest level since World War II. Statistics from the Aviation Safety Network revealed that there was a total of thirty-four deadly crashes around the globe involving multi-engine passenger planes. You'd have to go back sixty-five years, to 1946, to find a lower annual number of international flight tragedies. The total of thirty-four fatal accidents compared favourably to the average of forty-eight per year over the previous decade. Also, the total number of passengers and crew killed in accidents in 2001 – 1,118 overall – compared favourably to the previous decade's annual average – 1,298. Aviation can serve as a role model for other industries.

If we could transform flying from a game of airborne Russian roulette played only by courageous thrill-seekers to a largely uneventful mode of transportation used by millions of ordinary citizens, we may be able to improve safety and effectiveness just as successfully in other sectors.

World War II proved to be a turning point in aviation safety.[5] Urgent need fostered the invention of advanced, complex machines to support military activities, but of course, those machines didn't operate by themselves. Somebody had to use them, and during World War II, that somebody was usually a very fit, very young, very motivated male willing to die for his country. Yet despite the tremendous incentives, the reports that came back from the field, the sea and the air indicated that people were having serious problems with the new machines. The technology seemed to be too complex for the people using it. American Lieutenant-Colonel Paul Fitts, who was to play a key role in unravelling the mess that aviation was in at that time, summarized the situation, with military understatement: "The intense effort to produce new weapons, the race against time in industrial production, and the magnitude of the program required to train men to operate these new machines resulted inevitably in many instances in which the final man-machine combination failed to function effectively."[6]

In other words, far too many airplanes and crews were lost, not just during combat, but during training. The now-famous phrase "pilot error" began to appear in accident reports. The lieutenant-colonel and his colleague Captain Richard Jones captured the hair-raising threats to safety by using what has since become known as the *critical incident technique* – what I would characterize as a Human-tech methodology: professionals were asked to recall and describe particularly challenging incidents that were associated with an error.[7] Fitts and Jones interviewed or surveyed many pilots about their flying experiences during the war, and were able to collect over seven hundred "critical incidents."

The pilots' statements read like the design equivalent of a house of horrors. They are a long list of cockpit deficiencies indicating an incredibly bad fit between the pilots and the complex technology they were struggling to control. Many of the problems arose from mismatches or dysfunctional relationships between the physical shape, size or location of the cockpit controls and the size, shape and dexterity of the human body. In this critical incident report, a flight instructor describes how he confused one control for another during a crucial moment and almost lost his life:

> I started up my airplane, BT-13, taxied out to the take-off strip, ran through pre-take-off check and proceeded to advance the throttle. I held the plane down to pick up excess airspeed and as it left the ground, proceeded to pull back on the prop control to lower the rpm. Immediately, the engine cut out and I could see nothing but fence posts at the end of the field staring me in the face. Luckily, I immediately pushed the prop control forward to high rpm's and the engine caught just in time to keep from plowing into the ground. You guessed it. It wasn't the prop control at all. It was the mixture control.[8]

Another pilot unknowingly activated a switch because it was located too close to several other switches:

> Coming in on the final approach in a B-17, the pilot asked for landing lights. The flaps were one-half down and we were about 2,500 feet short of the runway. When reaching for the landing lights the flap switch was accidentally hit, knocking the flaps up causing the [aircraft] to mush into the ground. Major damage was done to the plane. I think this could have been prevented had the switches been further apart.[9]

The technology described in the vivid incident reports collected by Fitts and Jones clearly exhibits symptoms of a Cyclopean

Mechanistic world view. All the technical details were well attended to: not only did the planes get off the ground, they were even capable of some pretty fancy aerobatic manoeuvres. The problem was that the technology wasn't fit for human consumption. Designers didn't pay enough attention to the human factor. That's why Fitts and Jones put the term "pilot errors" in quotation marks, even in the titles of their reports. They wanted to draw attention to the fact that these errors were usually signs of cockpit design deficiencies, not personal failure. The strong link between inadequate design and "pilot error" was buttressed by one of Fitts and Jones's most compelling findings: almost all the pilots interviewed reported that they sometimes made errors in the cockpit, *regardless of their level of skill and experience.*

The reports authored by Fitts and Jones are now considered classics. They provide convincing evidence that aviation technology during World War II was in a state of transitional instability, wreaking havoc on equipment, pilots and their colleagues. The immense number of accidents, the tragic loss of life and the costly loss of equipment finally made the industry sit up and take notice. In step with Robert Wright's "logic of human destiny," social structures began to change as more and more designers realized the limitations of technical excellence alone. The cockpit defects that had gone unnoticed now stood out like a very sore thumb. And a group of pioneers – human engineers, as they were known back then – began to think about the problem in a new way, and came to the conclusion that the tragic and expensive loss of life and equipment could be drastically reduced by designing cockpits to fit the pilots' needs and capabilities.

One of the earliest and best-known success stories involved a recurring problem that had puzzled accident investigators.[10] After landing, the pilots of several kinds of military airplanes – B-17s, B-25s and P-47s – frequently retracted the landing-gear wheels rather than the flaps. You can imagine the effect: pulling up the wheels after you get an airplane on the ground isn't a very good idea. Why on earth would experienced pilots do this? It was as if a malicious "invisible hand" were in charge of the cockpit, shoving the airplane toward accidents.

In 1943, Alphonse Chapanis, a lieutenant in the U.S. Army, was asked to clear up the mystery. By adopting what was essentially a Human-tech perspective, Chapanis discovered that the root cause of these accidents wasn't the famous catch-all "pilot error," but rather technology that had no affinity with human manual dexterity. In the accident-prone aircraft, the switches or levers controlling the landing gear and the flaps were right beside each other and nearly identical in appearance. On a different aircraft – the C-47 – they weren't next to each other and were operated in distinct ways, so they weren't so easily confused. Not surprisingly, there were no landing-gear accidents with the C-47s – convincing evidence that "pilot error" could be prevented if the design of the cockpit controls was tailored to people's physical behaviour.

Moving the controls farther apart wasn't possible on existing aircraft, especially planes that were being used during the war. As a short-term fix, Chapanis modified the two controls by attaching a small wedge-shaped end to the flap control and a small rubberized disc to the wheel control. The differing shapes and textures made it easy for pilots to distinguish between the two controls merely by touch, and they were also easy to remember because there was a clear and intuitive relationship between the shape of the controls and the functions they controlled – a wedge-shaped control for the wedge-shaped wing flap and a round, rubber control for the round, rubber airplane wheel. And what impact did these quick fixes have? Pilots using the new design stopped retracting their wheels after landing. So-called "pilot error" was eliminated; as simple as that. The malicious "invisible hand" had been replaced – by design – with a benevolent "invisible hand" that effectively steered the airplane away from accidents. After the war, these easy-to-distinguish and

easy-to-remember controls were standardized on aircraft world-
wide – a lasting legacy from the seminal days of Human-tech
thinking. Chapanis's design was revolutionary at the time, but it
was subsequently generalized in the form of a design principle,
known as *shape coding*, that can be used in all kinds of situations
where it's important that controls be instantly identifiable.

These and other success stories fuelled a Human-tech
Revolution in military and commercial aviation. Fitts, Chapanis
and other pioneers showed the value of designing cockpits with a
human face. From then on, purely Mechanistic thinking was
largely a thing of the past in aviation, and partly as a result of this
conceptual revolution, flying eventually became, literally, hundreds of times safer than it was before World War II. The aviation
industry became *the* role model for Human-tech thinking.

Chapanis's elegant design change underscores the huge
impact that can be achieved by designing to fit the human body.
His shape-coded controls were an extremely small, simple modification – not only did they not require designers to scrap the
old design of the aircraft and start from scratch, they didn't even
require a change in the position of the two controls. But the
impact of this tiny design change was enormous: on countless
occasions it made the difference between an easy, safe, smooth
landing and the possibility of a catastrophic crash, saving aircraft
and pilots too. As we'll see throughout this book, this amplification effect is found at all levels of the Human-tech ladder, not
just the physical level.

THE ROCK 'N' ROLL REVOLUTION:
AN ELECTRIC GUITAR THAT FITS LIKE A GOOD SHIRT

The philosophy of fitting the design to the human body is infinitely transferable: the Human-tech approach can be applied to

a vast range of needs. One of the most fabulous examples was directly responsible for changing the course of musical history. Without the unique and powerful sound of the electric guitar, rock 'n' roll – the music of the Beatles, the Who, Led Zeppelin, the Rolling Stones, and Jimi Hendrix – just wouldn't exist. There's something raw and intense about the notes produced by an electric guitar that meshes perfectly with the rebellious strength behind the music itself.

But it was one particular electric guitar model that, arguably more than any other, influenced the revolution that became rock music – the Fender Stratocaster.[11] The Strat has been on the market since 1954, and the long list of famous guitarists who generated their influential trademark sounds on a Strat reads like a rock 'n' roll hall of fame: Buddy Holly, Buddy Guy, Jimi Hendrix, Richie Blackmore, Eric Clapton, Jeff Beck, Mark Knopfler, Stevie Ray Vaughan, Jimmie Vaughan, Robert Cray and Bonnie Raitt. The Strat was also behind the sound of some of rock's most popular and influential songs. Eric Clapton used a 1956 Strat – which he dubbed "Brownie" – on his hit recording "Layla." When that guitar was first manufactured, its list price was about US $250,[12] and Clapton bought Brownie second hand in 1967 in London for £150.[13] Because it was used to record a song that became one of rock's most famous love songs, the guitar's value skyrocketed. In 1999, Christie's of New York auctioned off Brownie on Clapton's behalf as part of a charity fund-raising effort. The selling price? I remember it well because I listened to the auction over the phone. I was (unsuccessfully) trying to buy one of Clapton's lesser-known guitars (there really is no such thing, as I eventually learned). If you include the commission fee, Brownie sold for a record US $497,500 (the case was thrown in for free). Half a million dollars for a forty-three-year-old guitar that had cigarette burns, scratches, and generally looked as if it had been run over by a Mack truck! You might think the buyer was completely out of his mind to pay that much money, but that was the going price on June 24, 1999, in New York City for a centrepiece of rock 'n' roll history. And the Fender Strat found itself literally in the spotlight.

The Strat was invented by Leo Fender, who was a radio repairman before he started building guitars. His company, Fender Musical Instruments, was created after World War II. During the war, when the efforts of previous musical manufacturers had been diverted to contribute to the wartime production of goods, it had been difficult for musicians to buy new instruments, and they were still scarce; also, guitarists in big bands were looking for a way to get their music heard over the sound created by louder instruments, like drums and horns. Fender believed there would be a demand for well-designed electric guitars, because their sound could be amplified, and hence heard even in a big band setting.

But rather than get embroiled tinkering with technical details as many Mechanistic-minded musical instrument designers did, and still do to this day, Fender had the vision to focus on the musician's needs from the beginning and to use those needs to steer his design effort. According to Freddie Tavares, Fender's colleague, "Leo Fender's general philosophy . . . was make it practical, as practical as possible and as simple as possible."[14]

Fender also had the wisdom to listen to his customers and run design concepts by them to get their feedback. As Tavares said, "One of the reasons for Leo's success was that all the musicians knew that they were welcome in our lab. They could come out there and talk to us directly. Everybody knew they could get to Leo."[15] He took the work out of the lab, and early on in the design process conducted field tests during studio sessions and live performances with professional musicians. Bill Carson was one of those musicians: "Leo had a really uncanny ability to take what you were telling him and interpret the needs of that as a [guitar] player. He didn't play guitar, he didn't think or hear in terms of a player, so he relied heavily on players."[16]

Fender repeatedly built prototypes, gave them to musicians to use, listened to their feedback and reshaped the design; then he would build another prototype and start the user-testing process all over again. The effort really paid off. A number of features that found their way into the final design of the Strat weren't present in the first prototype.

The overall shape of the body of the Fender Strat is one example. The bodies of previous electric guitars, squared off at right angles, were sharp and would dig into the guitarist's rib cage. Bill Carson told Fender about the problems he had experienced with another of Fender's guitars, the Telecaster: "The thing I didn't like about the Telecaster was the discomfort of it . . . I was doing a lot of studio work at the time on the West Coast and sitting down its square edges really dug into my ribs. . . . one of the things Leo got tired of hearing was that a guitar ought to fit you like a good shirt does."[17] Fender built prototype after prototype to improve the relationship between instrument and musician. Carson described the process: "I went one morning, this was early '53 as I remember, and Leo had sawed out four or five bodies . . . to see which one of those bodies did the job."[18] Thanks to many design iterations, the final version of the Strat has a comfortable, contoured body, a perfect fit. No more sore ribs.

Another design feature of the Strat was the receptacle where the cord plugged into the guitar; it was mounted on the front surface of the guitar within the musician's view, instead of at the bottom of the instrument, on the edge and out of the guitarist's view (its usual position on many other guitars). Fender players could easily see and reach the receptacle instead of blindly groping along the bottom edge of the body to find it. As Fender pointed out, "Another very important consideration was the position of the controls. On the Stratocaster we positioned them a lot nearer the guitarist's playing hand and that seemed very popular."[19] Also, traditional designs had three tuning pegs on the top and three on the underside of the guitar head, where they were out of view. Fender had the six in-line tuning pegs facing the player on the top of the guitar head, making it easier for the guitarist to grasp the peg for the desired string. As A. R. Duchossoir, author of *The Fender Stratocaster,* put it, "The purpose of Leo's invention was to make life easier for players."[20]

Other insights that he gleaned from guitarists after the Strat was released in the market eventually found their way into later models. For example, consider the design of the pickup selector switch.[21] Originally, Fender had designed this linear switch with three slots; it could be moved into three positions, allowing the guitarist to create three different kinds of guitar sound. Working musicians found they could actually create two new "out-of-phase" sounds by putting the pickup switch in intermediate positions: if you could get the switch stuck in between the first and second slots, you got a new fourth sound; similarly, if you could put it between the second and third slots, you would get a fifth sound. It took a fair amount of manual dexterity to put the pickup switch into these intermediate positions and keep it there – after all, the switch had been designed to have three slots, not five – but word spread amongst professional musicians that you could use this little trick to get some cool new sounds out of the Strat. Because Fender kept in touch with musicians, he learned of this trick and redesigned the pickup selector switch in 1977 to have five slots and thus five positions, so they could get the two out-of-phase sounds without having to engage in fancy tricks of manual dexterity.

Fender's genius led him to apply the Human-tech approach to the manufacturing and repair processes as well – which is where his experience as a radio repairman came in handy. He was sick and tired of trying to repair radios designed without consideration for the repair technician's job. "The design of everything we did," he said, "was intended to be easy to build and easy to repair. When I was in the repair business . . . I could see the shortcomings in the design completely disregarding the need for service. If a thing is easy to service, it is easy to build."[22] So the Strat was tailored to fit the needs and capabilities of the manufacturing workers and the repair technicians as well – an astoundingly enlightened design process for 1954.

The Fender Strat not only changed the course of musical history, but was also a tremendous success in the marketplace. We can get a feel for how unique the Strat is in this regard by comparing it with the other guitars that came and went. In the electric-guitar market graveyard, we find guitars whose labels read right side up from the audience's point of view, but upside down from the musician's point of view. Then there's the design that has over thirty controls! There are also designs that have unmarked multiple controls that look identical, but do completely different things. I could cite countless other examples of hopeless design features, but you get the picture. The Fender Stratocaster was an early example of wide-ranging Human-tech thinking, used in this instance in the service of artistic expression. Fender forged a direct connection between the guitarist's creativity and the sound coming out of the amplifier – the musician-and-guitar become one tightly integrated system. Poor designs, based on a Mechanistic world view, create a mismatch between the guitar and the musician, driving a wedge between the two and forcing the musician to concentrate on the knobs and dials rather than on the music. I'll give the incomparable Jimi Hendrix – a devoted Strat customer – the last word on the subject: "Music is getting better and better, but the idea is not to get as complicated as you can, but to get as much of yourself into it as you can."[23]

COMMERCIAL BLOCKBUSTER: THE REACH TOOTHBRUSH

Companies won't change the way they do business unless they can increase their market share while also improving customer satisfaction. The Fender Strat suggests those goals can be achieved by adopting Human-tech thinking. But is that really true, or was the success of the Strat just a fluke? Let's look at the design history of another commercial blockbuster – the Reach toothbrush. It provides an interesting answer.[24]

We all use toothbrushes, and there are many, many different kinds on the market. The traditional designs are plain, but there are plenty more with sophisticated features. They come in soft, medium and hard. Some have handles that are bent; others have straight handles. Some have rubbery grips on them; others don't. Some have flat bristles; others have bristles of uneven sizes.

Toothbrushes used to be pretty boring and pretty much the same, but that all changed in 1977 with the groundbreaking introduction of the Reach toothbrush. There are now many copycats on the market, but when it first came out the design was noticed because it had some unusual features. The toothbrush head was tilted at an angle, and it had a contoured handle.

The design of the Reach toothbrush was based on a human need. People tended to be pretty sloppy in brushing their teeth. Research showed that, on average, people spent only about a minute brushing their teeth – not enough time. Identifying that need – improved dental hygiene – provided DuPont de Neymours & Co., Inc. with an overarching direction, the functional equivalent of a rudder, that would steer all their subsequent design decisions.

DuPont hired a consulting company called Applied Ergonomics Corp. The design team included two ergonomists and one dentist. In the early stage of the design process, the team didn't focus on any particular technological solution. They reviewed how various designs prevented tooth decay and other dental problems. Their initial goal was to identify the most important problems in dental hygiene and to understand the nature of users' tasks. They learned that plaque removal was the most important goal and that gum massage was of only secondary importance. This insight helped them focus their design efforts;

any design concept that didn't help people do a good job of removing plaque shouldn't be considered.

During the next stage, a marketing questionnaire was sent to three hundred people to study their tooth-brushing habits. The feedback revealed that the inside surfaces of the back teeth were particularly difficult to reach with a traditional toothbrush. The strategies people use in brushing their teeth were also identified. Difficulty in brushing turned out to be a significant impediment to good hygiene. The more effort required, the more plaque would build up. The questionnaire also revealed the physical features that people look for in a toothbrush: a simple, plain handle, and a full brush head of level, white bristles.

Rather than rely on their own subjective opinions and preferences to make design decisions, the team gathered objective data about the size of people's hands, mouths and teeth. This information helped them determine the size and shape the toothbrush should be to fit comfortably in the hands and mouths of a wide range of people. The team also watched people brush their teeth and analyzed the different steps involved – a process called *task analysis* – noting details like how long people spent brushing various areas of their mouth, the way they manipulated the brush, and the direction of their brushstrokes. Almost everyone rotated their grip on the handle while they brushed, and half the people changed the position of their grip. These findings ruled out contoured or round grips because they would make hand movements difficult. The handle would have to be rectangular in cross-section to be easily grasped while the brush was being rotated or the grip repositioned. Further user testing and evaluation showed that a bristle size diameter of 0.27 of a millimetre turned out to be ideal for removing plaque, a fact that the design team didn't know before conducting the tests.

Two prototypes were constructed and tested to see which features were best at plaque removal, which would be best accepted by potential consumers, and how the new designs compared to several commercially available toothbrushes. There was no clear winner. Both prototypes tested better at plaque removal than the commercially available brushes, and the small head of one prototype was better because users could be more thorough, brushing only two teeth at a time rather than four or five. The angled, rectangular handle of the other prototype was better because it not only had a comfortable grip that was easy to grab and move, it also provided a natural area for a user to place his or her thumb, making brushing easier. The final product was a hybrid created by combining features from both of the two prototypes.

The result was a revolutionary innovation in dental hygiene. A deliberate and thorough attempt had been made to tailor the size and shape of the design to the size and shape of people's hands and mouths, all in the service of an important human need. The payoff was win-win: customers got a toothbrush that helped them maintain good dental hygiene and avoid expensive dental bills, and Dupont achieved a spectacular market success – the design is well known and has been commercially available since 1977.

HUMAN BODILY FUNCTIONS: THE FLY IN THE URINAL

The size and shape of the human body aren't the only physical aspects of human nature that tend to be overlooked by designers: basic facts about human physiology can also be ignored. I learned that lesson in Foley's class as well. He lectured us on a topic that first-year university students don't expect to hear about in typically staid engineering lectures – the design of a urinal. Exploiting the fact that his audience was primarily composed of nineteen-year-old males, Foley began by talking about the well-known problem of "splash back": when a man uses a urinal, the urine sometimes reflects off the porcelain and onto his pants, causing stains – not very user-friendly. Foley mused on how embarrassing this can be, particularly in summertime when you're wearing light-coloured pants.

We sat in stunned silence. Here was a university professor at the most prestigious engineering school in Canada lecturing to us about peeing on our pants! At the same time, we were very attentive because this was a pretty down-to-earth problem that all the men in the class had experienced and been embarrassed by (the women were intrigued too – and amused). Foley made points with scientific precision, showing us a graph of the trajectory that urine typically takes during urination. The curvature of the urinal is a critical design variable because it affects splash back. By this time we were all laughing, but we didn't lose sight of the practical implications: if the urinal design takes that typical trajectory into account, you can walk away with your pants unscathed.

Foley then showed how this solid engineering knowledge had been put to good use by some clever urinal designers. Beginning with basic human desire (to avoid splash back), then using knowledge of people (the typical trajectory) and of technology (how the curvature of the porcelain determines the reflection angle), the designers had adapted the technology to create a good fit between the two–a urinal whose curvature successfully minimized splash back. Good design, good hygiene. Very convenient. Particularly in summer. . . .

Years later, I learned of a creative design that took the idea of a user-friendly urinal one step further. If you go to the men's washrooms at the Schiphol airport in Amsterdam, you may notice there's a fly in the urinals.[25] So what do you think most men do? That's right, they aim at the fly when they urinate. They don't even think about it, and they don't need to read a user's manual; it's just an instinctive reaction. The interesting feature of these urinals is that they're deliberately designed to take advantage of this inherent human male tendency.

The fly isn't really a fly. It's a drawing of a fly, permanently etched onto the porcelain. And the etching isn't placed in just any old location on the urinal. On the contrary, it's been strategically etched into the "sweet spot" of the urinal, the point of curvature that minimizes splash back.

This design is even more elegant than the one that Foley described, because it draws on a whole range of understanding about people's needs and dispositions. The result is a seamless, harmonious relationship between people and technology that addresses a daily human need for basic hygiene. The idea makes so much sense that we wonder why it hasn't been applied more widely.

THE DARK SIDE: "USER-FRIENDLY" EXECUTION DEVICES

The idea of fitting the design to the body also has a dark side. The Human-tech approach begins with the identification of a human or social need, a problem worth solving, which is a value-laden decision: one person's need can be another person's undoing. The poster child to prove this point is Fred A. Leuchter Jr., an American engineer who made a career out of designing "inmate-friendly" execution devices that were exceptionally well tailored to the human body. This macabre story is told in Errol Morris's thought-provoking documentary film, *Mr. Death: The Rise and Fall of Fred A. Leuchter, Jr.* Leuchter's own deadpan description of one of his projects captures the tone far more revealingly than I ever could.

> A number of years ago, I was asked by a state to look at their electric chair. I was surprised at the condition of the equipment, and I indicated to them what changes should be made to bring the equipment up to the point of doing a humane execution.

Beyond making recommendations for changes, I sat down on my own time and at my own expense and made a new design and new equipment available to the states utilizing electrocution at a price far lower than they would have to deal with if they hired an engineering firm to redesign a specific item. The equipment is all standardized, it all meets the current electrical requirements for electrocution, and the pricing is such that it's similar to what you'd pay for an off-the shelf item. . . . They essentially paid for the parts, the labor, and the installation. And a 20 per cent markup – which is more than fair.

With chilling thoughtfulness, Leuchter went through the minute details of tailoring the design of his electrocution devices to the physiological properties of human nature: "Our electric chair contains a drip pan. All executees, during the execution, lose control of their bodily functions. . . . This is a disgusting thing when it occurs . . . and I think everybody knows that urine is highly conductive."

But Leuchter's design activities didn't stop at electrocution devices.

Because of my work in electrocution, I was contacted by the State of New Jersey to consult with them on the construction of a lethal injection machine. They realized that lethal injection is a difficult if not impossible problem, even for trained medical personnel. They determined that there should be some kind of a machine that could repetitively deliver the necessary chemicals at the proper time intervals for all executions. This completely took the human factor out of it.

It certainly did.

And as with his novel design for electrocution devices, Leuchter's creative ideas for lethal injection didn't stop at the technical details: "You certainly can make it more comfortable.

You could put him [the inmate] in a contoured chair, like they have in a dentist's office, that at least [sic] he'd be sitting up. You could give him a television. You could give him music. You could put some pictures on the wall. Rather than put him in a concrete room – that's not humane." Fred A. Leuchter Jr.'s, grimly considerate Human-tech thinking is a reminder that technology itself is value-neutral, and that this last century has seen the horrendous justification and creation of new technologies whose sole purpose is to kill people effectively, efficiently and "humanely." Beware the dark side of the Human-tech force.

The physical level of the Human-tech ladder provides a very concrete, and tangible understanding of the people-technology relationship. But as our society becomes more and more dominated by the information economy, the work we do increasingly draws on our knowledge rather than our physical strength or dexterity. It's not that physical considerations no longer matter, as the prevalence of repetitive strain injuries among people who work frequently with computers clearly shows, but rather that we're demanding more of our psychological capabilities. In the past, many jobs involved mostly manual labour and required us to exercise our muscles, not our brains. The human factor at the physical level continues to be of fundamental importance, but the nature of work is changing, which means that our technology must mesh with our human needs at more complex, sophisticated levels. So it's crucial that we move up the ladder and tailor products and systems not just to our bodies, but to our minds as well.

4

Minding the Mind I: Everyday Psychology

THE HUMAN-TECH LADDER

TECHNOLOGY (hard or soft)
Information content/
structure, cause/effect
relations

↺ NEED ↺

HUMAN FACTOR

PSYCHOLOGICAL

The second level of our ladder – the *psychological* level – introduces a host of new data about human nature that Human-tech design must take into account if it's to lead to fluent, productive interaction between people and technology. A particular technology may be well suited to our physical abilities, but products or systems that pose too much of a burden on our limited memory capacity, that are too complex or counter-intuitive for us to understand, or that violate our mental expectations about cause-effect relationships will fail to pass Robert Wright's test of "affinity with human nature." Other products or systems, those that are tailored to these psychological considerations, create a close bond between person and technology, and lead to a single, tightly integrated system, which can be so effective that the technology itself becomes transparent. Like a blind man with a cane, we don't have to think explicitly about the technology, and are free to focus on pursuing our goals.

This is where my own discipline, human factors engineering, comes most into play. One of the things my colleagues and I do is document the psychological properties of people and the design techniques that can be used to create a fit with those properties. The best-known book on the subject is Don Norman's bestseller, *The Psychology of Everyday Things*.[1] In chapter 5, I hope to show how "minding the mind" can facilitate the operation of safety-critical systems – health-care systems, airport security, and so on – but I want first to prepare the ground by discussing something far less threatening: the familiar products and systems that we use to perform everyday tasks (for ourselves or in our offices and businesses) like shopping, driving, entertaining ourselves, cooking, or managing our time.

"YOUR CALL IS IMPORTANT TO US":
THE LABYRINTH OF AUTOMATED PHONE MESSAGE SYSTEMS

Human psychology is particularly important in the design of everyday products and systems because many of the electronic products we currently use fit our bodies but not our minds. My favourite is the omnipresent but dreaded automated phone message system, which has all but replaced human operators or receptionists. Once you receive your directions and set off in search of the information you want, manually pressing the keys on the telephone is simple enough; they're easy to reach, easy to depress, and easy to read. The technology has a good fit at the physical level. No problem there.

At the psychological level, it's a whole different ballgame. Half the time, we can't figure out which magical sequence of button presses will take us through the byzantine set of options to the relevant function. We start off by listening to all the available selections listed in the "main menu," and then we pick the one that seems to best match our goal – but when we get to the next layer of menu options, we belatedly realize that we've gone down the wrong path, and none of the new options is relevant to our goal. If we're lucky, we remember which button to press to get back to the "main menu," or maybe we just hang up. In either case, we have to start all over again. Far too often, I've persistently gone through this maddening process several times during the same call, until I finally figured out the right sequence. All we're trying to do is purchase a product or service, or ask a simple question. It should be quick and easy, but it rarely is.

I don't think the Wizards who design these systems really understand the infuriating impact their creations have on our daily lives. Recently, I used an automated phone message system that gave me the option of what kind of music to listen to while I was on hold. "Press 1 for easy listening, press 2 for country, press 3 for jazz, . . ." Is this supposed to be the answer to our frustrations? The problem isn't what kind of music we have to listen to; it's the fact that the technology shows a complete lack of respect for human beings: there are too many menu options for us to remember, it's hard to understand which of the available options is the one we need, there are too many layers of menus for us to keep track of where we are in the hierarchy; and it takes far too long to get the job done. There are times when we've probably all felt like screaming into the phone receiver.

Nevertheless, automated message systems are technically sound – the menu of options is stable and reliable; when we press a particular button to select one of the available options, we're always presented with the same message. Whoever designed the hardware and software knew what they were doing from a strictly technical perspective. Ralph Nader, consumer advocate and former U.S. presidential candidate, told a rather amusing anecdote at a conference I attended. After summarizing the exasperating problems created by the widespread introduction of automated phone message systems, Nader sheepishly told us, "Sometimes when I'm in the office working late at night and want to listen to classical music, I phone up United Airlines and use my speaker phone to listen to the music they play while I'm being put on hold." Nader's little story offers evidence of the technical dependability of the design – that's not where the problem lies at all – but it reveals a classic symptom of the Mechanistic world view: the technology is sound, but not enough attention has been paid to the human factor if we're put on hold for a long time. There's a bad fit there between us and technology.

In many cases, difficulties are created because companies use their internal corporate structure as a basis for organizing the options listed in the menus, rather than organizing those options in ways that make sense to the consumer. For example, in some banks, the credit-card division is organizationally distinct from the division that deals with chequing and savings accounts, which in turn, is organizationally distinct from the division that deals with mutual funds. As a result, their automated phone message systems are sometimes designed so that you have to follow very different paths, depending on which organizational unit your request falls under. But as consumers, we don't know – and shouldn't need to know – what our bank's organizational structure looks like. So if we want to move some money from our chequing account to a mutual funds account (so that we can actually earn some interest on our money), there may not be a simple way to do this because our request cuts across organizational divisions. Instead, we may have to engage in a long sequence of actions to traverse several branches in the menu options for the chequing account division and another set of branches in the menu options for the mutual funds division.

This is a complete waste of our time. The menu options should be organized in ways that fit the consumer's needs and expectations, not the bank's.

GOING FOR A RELAXING DRIVE: THE MAGIC ROUNDABOUT

Sadly, navigating the electronic maze of telephonic paths isn't the only psychological challenge imposed by everyday technologies. Sometimes, navigating in the real world can also turn into an adventure, even for those who aren't spatially challenged. For instance, if you happen to be driving near Swindon in England, you may find yourself faced with a sign informing you that you're about to enter the "Magic Roundabout." Had you signed up for a Magical Mystery Tour, you might expect such a sign, but most out-of-town travellers are likely to be surprised.

The traffic sign for the Magic Roundabout in Swindon, UK

If you've travelled in Europe before, you're probably already familiar with the concept of a roundabout. Instead of using traffic signals at intersections to control traffic, a roundabout provides a circular path for regulating traffic entering an intersection of several streets. The circulating traffic has the right of way, so as an incoming driver you're supposed to merge seamlessly into the roundabout when you see a gap in the circulating traffic. You drive around the circle until you find the exit you're looking for and then peel off.

Traditional (non-magic) roundabouts can present a challenge, especially during rush hour. (I've been in cars with timid drivers who've just kept circling around and around because there was a lot of traffic and they were stuck on the inner lane and didn't have the nerve to make a break for the exits on the outer part of the circle.) They're already complex and demand attention and mental agility. But as its name indicates, the Magic Roundabout is no ordinary roundabout.

If you look closely at the photo on the previous page, you'll see that the Magic Roundabout is actually a roundabout of roundabouts. The larger circle is made up of five smaller circles that are connected together. Each of those smaller circles is itself a roundabout. The implications for the driver are easier to appreciate if we look at the bird's-eye view of the Magic Roundabout shown in the next photo. Note that the asphalt is smooth and the paint marks on the ground conspicuous; there are no potholes caused by wear and tear and the arrows haven't yet been erased by the rainy British weather. It's good that these technical details have been taken care of, but I probably don't need to describe what it would feel like to drive through this maze of circles, which is vaguely reminiscent of the maze of options provided by automated phone message systems. A mere glance at the photo is enough to make your brain tired – the mean-spirited nature of the "invisible hand" in this design is palpable.

A bird's-eye view of the Magic Roundabout.

But people are smart and ingenious, and so it turns out that the Magic Roundabout is actually not as challenging or as dangerous as it appears. The Swindon County web site concedes that the Magic Roundabout "may confuse or amuse new visitors and baffle American tourists," but it goes on to point out that "there have been [only] 14 serious accidents and 59 lesser ones recorded in 25 years."[2] I'd suggest these statistics are a remarkable testament to human adaptability. Despite the lack of thought given by the designers to our navigational capabilities, the vast majority of drivers are apparently able to weave their way safely through the labyrinth of the Magic Roundabout.

"HEY, BILLY, WHAT TIME IS IT IN YOUR HOUSE?": REMEMBERING INSTRUCTIONS

Other everyday products handicap us with the demands they impose on our memory. Let me give you a personal example. When my whole family gets together for a meal over the holidays or for someone's birthday, being Portuguese, we drink wine, eat quite a bit, and talk even more. Meals can last several hours. When these visits took place at my aunt and late uncle's home we would sit in the living room after dinner and watch a movie. But watching TV there was distracting. However hard I tried to focus my attention on the screen, it would still be diverted: the VCR clock was always flashing "12:00."

I used to reset the clock for my uncle. I would show him the steps, and he would nod and seem to get it. But, inevitably, when I visited him a few months later, the clock would be flashing again.

Of course, it was *possible* to set the clock on the VCR. The electronic circuits behind the sleek facade contained the proper components, with the required voltages. And my uncle was a smart person – certainly his intelligence wasn't the problem. It was just that the sequence of actions required to perform the task was too counterintuitive for him to remember. My uncle's difficulties were entirely predictable, because one of the things we know about human memory is that it's difficult to recall things that aren't meaningful to us. If he had to set the clock every day, he would eventually have learned the proper sequence of actions even though it was counterintuitive, but since he didn't need to set the clock very often, he just gave up.

My uncle wasn't alone. Difficulties with VCRs are so common they've now been immortalized in a television commercial that shows a group of kids playing outside. One kid says to the other, "Hey, Billy, what time is it in your house?" Billy holds up both his hands and alternates between completely extending his fingers and closing them all the way to form two fists, manually mimicking the flashing indications on his parents' VCR while simultaneously replying to his friend's question: "Twelve o'clock, twelve o'clock, twelve o'clock."

A similar memory collapse occurs every six months, when the time changes, and many people struggle to remember how to advance or turn back the time on the clocks in their cars. (Some of us actually do this while we're driving, posing a hazard to others on the road.)

SMALL CHANGES, BIG EFFECTS:
THE FOUR-BURNER PROBLEM

Another consideration at the psychological level is the intuitive expectations we all have about cause-effect relationships. This sense of what-ought-to-be-where can help us to make sense of and interact with the world, but many everyday products don't take this into account. Take the ordinary stove, for instance. Figure 1 shows a typical design solution to the so-called "four-burner problem": how to lay out the knobs to control the four burners on the stovetop. Simple? Not so simple. The four knobs, shown at the bottom of the diagram, are arranged in a linear sequence, but the four burners aren't. They're laid out in a square arrangement with one burner in each quadrant. (I've added the labels A, B, C and D so that you can see which knob controls which burner on the stovetop.) Designs of this type can be found in millions of households, but just because a design is popular doesn't mean that it's easy for people to use.

Granted, the correct controls are easy for us to reach in physical terms, and the stove is perfectly capable of cooking our food – the wiring inside and the burners on the stovetop are technically flawless. But the design isn't easy to grasp psychologically

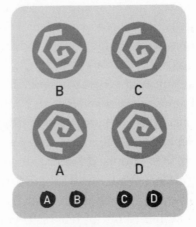

Figure 1. A typical stove design showing the location of four burners and their respective controls. The labels A, B, C, D are not part of the stove's design, but have been added by me to show which knob controls which burner.

because the relationship between the controls and the burners violates our expectations. We typically expect that controls that are beside each other – A and B, for instance – would control burners that are *also* beside each other. This tendency to use logical cause-effect relationships when interacting with the world around us is about as close as psychology comes to an immutable law of nature; the principles we use to infer such relationships represent a remarkably consistent fact about human behaviour that is extremely difficult to change. Yet this common stove layout violates these basic expectations. The design wasn't created with the human factor in mind.

We've all occasionally grabbed the wrong knob on stoves of this type. I've met several people with graduate degrees who've confessed to me that they've burned a kettle or two because they used the wrong control, and unintentionally turned on a burner that had an empty kettle on it. Sometimes we notice our mistake only minutes later when we smell the smoke coming from the melted plastic handle on the kettle. Usually, we blame ourselves. Shouldn't we be able to grab the right control every time? After all, it's a stove, not a nuclear power plant.

But what makes this type of problem incredibly frustrating to Human-tech thinkers is that a better design has been available since 1959. That's right, *1959*. Again, we have the remarkable Alphonse Chapanis to thank – he was the army lieutenant-colonel who fixed the problem with the confusing cockpit controls in World War II.[3] Rather than try to change human nature and expect people to conform to technology, Chapanis studied how the layout of our stove controls could be adapted to our human expectations.

The design he came up with is shown in figure 2. Only one simple change has been made to the error-inducing design shown in figure 1: the burners have been offset slightly so that they're ordered in a linear sequence from left to right (A through D). Now there's an obvious, intuitive relationship between the

positions of the knobs and the
positions of the burners. The leftmost knob controls the leftmost
burner. The third knob from the
left controls the third burner
from the left. And so on. The
general design principle referred
to here is known as *stimulus-
response compatibility*, and it can
be used to ensure a psychological affinity with human nature
in all kinds of products or systems that involve cause-effect
relationships.

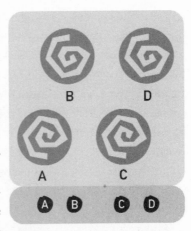

Figure 2. A Human-tech solution to the four-burner problem.

The impact of this design change is the same as that of
Chapanis's cockpit design change. Experiments showed that
errors went to zero. *Zero* – no mistakes whatsoever. It's completely obvious how the device is supposed to be used – the
design acts as a helpful, not a mean-spirited, "invisible hand." No
more bad fit. And notice that going from an error-prone to an
error-impervious design required only a small, simple change.

WHICH WAY IS UP? CAR KEYS AND BATTERIES

Chapanis's use of the stimulus-response compatibility principle
wasn't a rare fluke. Human-tech thinkers have developed a host
of systematic design principles that take the human factor into
account. One of these principles, known as the principle of
behaviour-shaping constraints, is essential to designing usable
technology into products or systems. The basic idea is that
every object has limited possibilities for action. You can do some
things with it, but not others, and so the object can be said to

shape behaviour – although it doesn't determine it uniquely, because several actions are usually possible. For example, a chair is designed so that you can sit on it. It can also be used for other purposes it wasn't designed for. If you're painting your ceiling and you don't have a stepladder around, then a chair can also be used to stand on. If you're out in the woods and freezing to death, a wooden chair can even be used to make a bonfire to create warmth. However, there are things you can't do with a normal chair, things that just won't work – no matter how hard you try, you can't use it to fly. Granted, the example is extreme, but it makes the point: objects shape, or constrain, people's behaviour.

The goal of Human-tech thinking should be to design objects so that their behaviour-shaping constraints are likely to produce desirable outcomes – helpful "invisible hands." At the same time, objects should be designed so that they don't have behaviour-shaping constraints that lead to undesirable outcomes.

My Volkswagen GTI car key "shapes" my behaviour in a very elegant way. It was designed so that it's impossible to insert it into the lock in the wrong direction. How is this possible? It's not because of some Mechanistic solution, like fancy sensor technology. It's just that the key is symmetrical, so as long as it's aligned with the lock it will go in, regardless of whether it's the "right way" up or not. In fact, the terms "right" and "wrong" way up or down, which can be used to describe the way you'd hold a traditional key, don't make sense for this design. There's no such thing as upside down because the key-lock system has an easy-to-perceive, and easy-to-use, behaviour-shaping constraint that leads to a desirable outcome – a natural, seamless fit between me and my car key. The relationship is invisible – I can't grab it – but it's always there waiting for me and at work each time I use my car. I don't even have to think about what to do; I just put the key into the lock.

But how many times have you struggled to put a new set of batteries into an electronic toy or your remote control, only to find that you've put them in backwards and the thing won't function? Most batteries for small devices have a cylindrical shape. Because the battery is symmetrical, there seem to be two correct ways of inserting the battery into the device. However, if it's not correctly oriented the polarity will be reversed and the device won't function. The symmetry of this design, unlike the symmetrical design of my car key, is an undesirable feature, because it makes it just as easy to insert a battery the wrong way as it does to put it in the right way. The design complicates what should be a simple task by making it more likely that we'll respond incorrectly – that we'll engage, in effect, in undesired behaviours.

Happily, the battery for the laptop that I'm using to write these words doesn't suffer from this problem. The designers were thoughtful enough to construct an asymmetrical battery with several notches in it; it will physically fit into the laptop in only one orientation – the one that makes the computer work. As with my car key, it's virtually impossible to do the wrong thing and very easy to do the right thing, showing both the power and the value of behaviour-shaping constraints.

THE POWER OF FEEDBACK: ZIP-LOCK BAGS

Another powerful design principle that reflects Human-tech thinking is to provide people with immediate, salient *feedback* (or information). Fire is a simple yet powerful natural example that illustrates how this design principle affects our behaviour. We don't have to get burned to figure out that a campfire, for instance, can hurt us. As we approach the fire, we can feel the temperature increasing. Backing up provides a welcome source of relief, whereas even a single step forward can provide an immediate warning not to go any farther.

As we'll see throughout the book, the feedback design principle can be applied in all kinds of situations at almost every level of our relationship with technology. At the psychological level, it has many uses – like the design of zip-lock plastic bags. Yes, the kind that you put sandwiches in or that you use to store leftovers in your freezer. With the older designs, it was difficult to know if you had zipped up the bag correctly to create an airtight seal. There was no obvious feedback to tell you whether you had achieved your goal. At least, not right away. If you took the bag out of the fridge a week later and found that its contents had turned into a mouldy ecosystem because the bag wasn't sealed properly, the feedback came too late to do you any good. But some new designs use a clever colour-coding scheme to help consumers seal the bag correctly. One side of the zip lock has a yellow strip, and the other side has a blue strip. When you've squeezed the two sides together correctly, the two colours blend into a green line, signalling that the bag has been sealed. If the two sides haven't meshed correctly, you still see the two individual colours. With this design, it's more difficult to make a mistake because information is provided to allow us to discriminate successful from unsuccessful actions: green is good, yellow and blue are bad.

The existence of these general design principles is crucial because it means we already have the knowledge to design better products and systems in all kinds of areas. But broad change rarely comes about unless it's driven by the expectation of economic and business gain. Will Human-tech products designed to fit with consumers at the psychological level also lead to increased market share and improved customer satisfaction, as they did at the physical level with the Fender Strat and the Reach toothbrush?

HUMAN-TECH DESIGN SELLS: THE PALMPILOT

I hang out with people who always have a lot on their plate. With many demands on their time, they need to keep track of a lot of information, and so have a strong incentive to schedule and prioritize their tasks. They used to carry around black books – or sometimes even big binders – to schedule their time and to record important information. More and more, they've started turning to personal digital assistants (PDAs) – those little hand-held, computer-based devices that are the electronic equivalent of an appointment book. Interestingly, most of these people started using the very same model – the PalmPilot. How did one company achieve such deep market penetration in so little time? It's not because they were first on the market. That distinction belongs to Apple, but its Newton model was a complete flop in the marketplace. So what did Palm do to be so successful?

The designers started off by identifying an unsatisfied need in the marketplace, a rudder that could be used to steer the design process. First-generation PDAs hadn't done very well, because, as Rob Haitani, product manager of the original Pilot, pointed out, "the hardware vendors didn't understand the basic problem. They thought the first-generation products failed because they did not provide enough functionality."[4] This is a classic Mechanistic mistake; the technology Wizards ignored human needs and capabilities by loading down the product with as many features as they could think of. Designers at Palm had a different hunch. They thought that people would want a device that was small, fast, convenient and *easy to use*. It had to have the basic, essential functions that almost everyone would want, and very little else.

Rather than using their personal opinions to guide the design process, the Palm designers conducted market research to identify the basic functionality that people were interested in and would

use. Over and over, users told them that less is more. "I don't need
the bells and whistles. I just want to organize my phone numbers
and my schedule. Give me something that does that then hooks
up with a PC very elegantly, and that's great."[5] So while most other
companies were still trying to cram in all the features that were
technologically feasible, Palm listened to users and learned what
they really wanted: address book, date book, calculator, memo pad
and To Do list. Five functions, not eight hundred; pretty different
from the BMW 7 Series that I described earlier.

User research also helped Palm designers learn another valuable – but initially counterintuitive – lesson that had a big impact
on the final design. First-generation PDAs were designed to free
people from their personal computers, but Palm's user research
showed that 90 per cent of their customers owned PCs, designers thought it would be useful to offer them a data link from the
PalmPilot to the PC so they could synchronize the information
on the PDA and the PC. It turned out this was a function that
people would pay to have, but once again, they said it had to be
simple. Palm came up with an elegant design solution: all the
users had to do was press one button. A no-brainer.

In the detailed design phase, the Palm designers created prototypes to test their ideas and refine their design concepts. Jeff
Hawkins, the founder of Palm, got the user-testing process
rolling. According to Rob Haitani:

> Jeff believed we had to make the product considerably smaller
> than current PDAs. He carved up a piece of wood in his
> garage and said this is the size he wanted. He'd walk around
> with this block in his pocket to feel what it was like. I would
> print up some screenshots as we were developing [the user
> interface], and he'd hold it and pretend he was entering things,
> and people thought he was weird. He'd be in a meeting furi-
> ously scribbling on this mockup, and people would say, "Uh,
> Jeff, that's a piece of wood."[6]

Despite the primitive nature of the wooden prototype, it gave designers a good idea of which features matched users' needs and which ones did not. Rob Haitani describes what it felt like to use the final product, a benevolent design where the person and technology become one:

> Think about how you feel about PC applications. I swear at my PC all the time and I know of others who do as well. It gets very frustrating. You get angry and irritated when you can't do what you want to do. But if you apply the less-is-more approach, your [program] just does the right thing. For example, if you're in a hurry and you pull out your Palm and you want to know where you're supposed to be. You press one button and it shows you your schedule. No struggle, no effort, no confusion; it just does what you want to do. Your reaction is a contented sigh, and it's very calming. Your blood pressure doesn't rise. That's the key to the whole experience.[7]

How often do we have an electronic consumer product that could be accurately described as calming?

Palm also did a great job at the sales phase. They set up a discussion board on their web site where customers could share ideas.[8] There was also a "Cool User Story" section where customers could send in messages about their experiences with Palm's products. These efforts at keeping in touch with customers served several purposes. The customers felt as though the company was interested in their opinions, which created brand loyalty and customer acceptance. The cool user stories also provided invaluable information for marketing campaigns to show how effective the product was and how well it had been received. Finally, by obtaining customer feedback, Palm was able to identify new ideas that could serve as the basis for their next generation of products. This strategy worked very well. After its initial blockbuster success, Palm started offering a number of different PDAs as well as software and add-ons designed partly in response to comments received from users of earlier-generation products.

The Palm case study reinforces the lesson: Human-tech thinking isn't just good common sense, it's also good business sense. Companies can't use arguments about excessive cost or lack of market impact as excuses for not heeding human psychology when they design their products or systems. On the contrary, the evidence indicates that the Human-tech approach will advance, not impede, corporate objectives.

THE DARK SIDE REVISITED: DECEPTION AND MANIPULATION

The idea of designing products to fit the mind has tremendous untapped potential, but here too there is a dark side that we should be wary of. Figure 3 provides a simple, seemingly innocent, but exceptionally seductive example.[9]

Consider the vertical distance between the two curved lines on the graph. As they move toward the right, are they:

a) getting closer together;

b) getting farther apart;

c) a constant distance apart?

If you're like most people, you're surprised and may not even believe that the correct answer is that they're a constant distance apart. You'd swear that the two lines are getting closer together. Try it for yourself; get a ruler out and measure the vertical distance between the two exponentially-increasing lines. They really are a constant distance apart. In fact, I constructed the bottom line by copying the top one, pasting in the copy, and then shifting it vertically by a short distance.

The two lines seem to be converging because our visual system didn't evolve to measure the vertical distance between two lines accurately. Our eyes tend to focus on the nearest distance

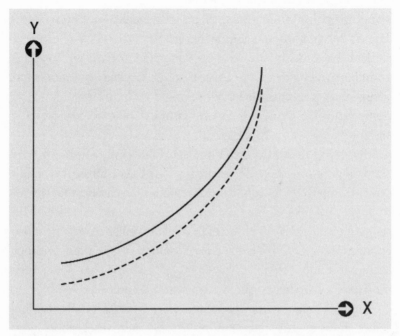

Figure 3. An example of how knowledge of human capabilities and limitations can be used to deliberately deceive.

separating the two lines, so as the curves bend upward together, the shortest distance between them gets smaller. However, statistical graphs are intended to convey information not according to the shortest distance between the lines, but rather according to vertical distance (represented by the y axis) and horizontal distance (represented by the x axis). That's why the graph can be misleading.

I mention this example to show that if you know something about human psychological capabilities and limitations – in this case, that our eyes tend to measure the nearest distance separating the two lines – you can deliberately exploit that knowledge to deceive people – in this case, design a statistical graph that creates a very compelling but false impression. Moreover, the person who's looking at the graph won't even know they're being misled by the pointing finger of a sneaky "invisible hand." The fit between the design of the graph and our psychological capabilities and limitations is so snug that the deception goes undetected. That's good news for the deceiver, but not such good news for the deceived. Human-tech thinking, and its tremendous influence on human behaviour, can be co-opted for all kinds of sly purposes.

"Shopping scientists" such as Paco Underhill, author of *Why We Buy: The Science of Shopping*, have already started travelling down this path.[10] Underhill and his band of researchers follow up to fifty shoppers a day as they move around a store. The researchers are skilfully discreet in their surveillance, getting close enough to see exactly what you're looking at or doing, without letting you know that you're being watched and that your every movement is being recorded with scientific precision.

And the goal of all of this arduous spy work? To better understand shopper-store interactions. As a result, researchers now know a great deal about how we shop. And what do the shopping scientists do with all this hard-earned knowledge? Supposedly, the aim is to put these insights to practical use "to make stores and products more amenable to shoppers."[11] This strategy sounds like Human-tech thinking; the store and its infrastructure are the technology, the shoppers are the people, and the design aim is to create a seamless fit between the two by adapting the design of the store to the psychological characteristics of the shoppers. Indeed, one of the tenets of the science of shopping places the human factor front and centre: "There are certain . . . abilities, tendencies, limitations and needs common to all people, and the retail environment must be tailored to these characteristics."[12]

But what's the human or societal need that's being fulfilled? Some of Paco Underhill's recommendations are definitely aimed at making life easier for shoppers – putting products that are purchased by the elderly in a more accessible location is an example. Nothing sinister about that. But other recommendations seem to be aimed at persuading people to buy, buy, buy – making life better for the merchants, not the shoppers.

Consider the "interception rate" – the proportion of customers who go into a store and have some kind of contact with a salesperson. Apparently, research shows that increasing the interception rate increases the amount the average shopper spends. Why? Because "talking with an employee has a way of drawing a customer in closer."[13] So the science of shopping recommends that stores have plenty of employees around to achieve a high interception rate and thus higher profits.

Now consider the subtle design of store signs. A study showed that a computer-based display with moving text was read by 48 per cent of the customers, whereas an earlier low-tech display with static text was read by only 17 per cent of the customers. The recommendation? Adopt attention-grabbing dynamic displays to convince customers to step in and buy.

One more shady example. Some supermarket stores provide painted-on graphics for a hopscotch game on the floor right in front of the shelves stocked with breakfast cereal. The results were impressive: "The average time kids played games in the aisle was almost fourteen seconds–a long time to be standing in front of the cereal without buying some."[14] The science of shopping doesn't just take aim at adults.

Does all of this add up to making "stores and products more amenable to shoppers"? You decide for yourself. Imagine this scenario as a test case: you and your kids stroll into a store that's loaded with "interceptors" whose sole goal is to "draw you in closer" until you're captivated by the flashing, moving-text signs. They in turn lure you straight to the cereal section, where your kids will keep on playing hopscotch until you give in and buy them several boxes of their favourite sugar-coated cereal that has a marketing tie-in with the latest Disney movie. I don't know about you, but I wouldn't call this a good shopping experience.

It's more like I'm being waylaid by beckoning "invisible hands" all around me – a system deliberately designed to lure me in, take advantage of my impressionable human psyche, get me to spend more money than I intended, and rack up profits for the retailers. And, as with the statistical graph example, I likely won't even know I've been duped. *Caveat emptor.*

Such is the state of affairs with our familiar, everyday interactions with technologies – millions of gadgets, products and devices that sometimes help us, but more often than not irritate and confuse us and – more than we admit or even know – deceive us. But the design of technology plays a crucial, if less commonly appreciated role, in more complex systems like aviation, airport security, the environment, nuclear power, and health care. The consequences of failing to heed human nature at the psychological level are far more severe in these safety-critical sectors. We would hate to think that complex technological systems are conceived in the same spirit of reckless innovation that dreamed up the Magic Roundabout, for instance. That kind of irresponsibility would invite airplane crashes or hijackings that could kill hundreds of people in a split second, environmental damage that would annihilate our planet, nuclear meltdowns that would spew radiation into the atmosphere, or medical catastrophes that would maim or kill patients in hospitals. Given the enormity of what is at stake, we can't possibly have the same kinds of difficulties controlling our complex technological systems that we have in the comparatively mundane everyday situations I've described here. Surely in safety-critical systems, where the consequences of a bad fit could destroy not only life but also the natural environment, extraordinary care has been taken to ensure that the technology is within human reach and therefore well under our control.

Or has it?

5

Minding the Mind II: Safety-Critical Psychology

In 1992, I arrived at Lester B. Pearson Airport in Toronto on a hot and muggy August day after yet another business trip. I was tired and just wanted to get home. After I cleared immigration, I had the misfortune of being sent to that section of customs where you get interrogated, and your luggage is opened and inspected. After waiting impatiently in line for what seemed like forever, I was finally interviewed by a customs officer. Our conversation went something like this:

Customs Officer (CO): Where are you coming from?

Me: Japan.

CO: What were you doing there?

Me: I was visiting the Japan Atomic Energy Research Institute.

CO: Why were you doing that?

Me: Because they invited me.

CO: Why did they invite you?

Me: I guess because of the work that I do.

CO: What kind of work do you do?

Me: I'm an engineering professor at the University of Toronto, and I do research on human factors.

CO (with a puzzled look on his face): Why would the Japan Atomic Energy Research Institute be interested in . . . that?

Me: Because I try to figure out how nuclear power plant control rooms can be designed to make it easier for people to do their jobs, thereby reducing human error and the possibility of a nuclear accident.

CO: That's interesting. So the idea is to make the control rooms more Homer Simpson–proof?

Me: Yes, something like that.

Because of the success of the American TV show *The Simpsons,* many people have a fanciful vision of nuclear power plants. Most of us don't make any connection between the technology we encounter in our daily lives – zip-lock bags, grocery-store displays and the like – and safety-critical, complex systems such as nuclear power plants or hospital operating rooms. After all, very few of us have any direct contact with such systems so we don't know what factors make them work or what might cause them to fail; they're so remote from everyday widgets and and worries that they might be on a different plane of existence. The design of zip-lock bags doesn't appear to have anything at all to do with saving the environment, and the design of batteries seems very remote from the formidable task of improving our safety when we check into a hospital.

Environmental issues are a good way of exploring this transition from everyday technologies to complex systems, because many of the tasks that we perform (or don't perform) on a daily basis – driving, using our computers, our photocopiers, disposing of our garbage – have a cumulative impact on the environment that can, slowly but surely, threaten the long-term viability of human life on planet Earth. So I'll begin with technology that has an impact on almost all of us these days – the ubiquitous PC – and move on up. Can Human-tech design principles used to create more user-friendly everyday technologies also be used to design "green" products and systems that are more environmentally friendly?

THE POWER PIG: CONSERVING ELECTRICITY ON YOUR PC

In 1994, I was teaching an introductory engineering design class to second-year undergraduate students and I was tired of following the same old routine, so I decided to try something different, something more ambitious, and something that, I hoped, would show the students how the engineering profession could make a meaningful difference in the world. For the class project, I asked them to apply a Human-tech approach to a global problem of their choice.

At first, I thought I had made a huge mistake and that the class was going to turn into a disaster. The students were confused and frustrated. How, they protested, were they supposed to complete an assignment like that? I was asking them to choose an overwhelming global problem that we're all familiar with – like non-renewable energy supplies, non-renewable natural resources, pollution or global warming – and address it, write up their idea in a report, and present it to the class – all in one semester. No wonder they didn't know where to start.

In an attempt to salvage the course, I tried to lessen their anxiety by pointing out that they didn't have to cure all of the world's problems in one fell swoop; just design a simple product that could make a modest dent in reducing a significant global problem. The key was to pick an environmentally unfriendly activity that was performed many times by many people and focus on the social aspects of the technology it employed. If they could design a product that would lead to a small social improvement, and that product was used frequently, then the benefits to the environment, and thus to quality of life, could slowly but surely add up over time. A tonne of feathers still weighs a tonne.

Much to my relief the students caught on. The ideas that they came up with were astoundingly innovative and show what strides can be made by designing products to fit with human beings at the psychological level.[1] The first example, developed by Dave Kuk, Jon Cowley and Fred Beserve, is my favourite.[2] They decided they would look for a way to reduce the amount of electrical energy used by desktop computers. This was a case of a small problem of social significance looking for a solution, rather than fancy Mechanistic technology with lots of features looking for a problem.

We might think that the amount of energy used by computers is tiny, but that's not the case. At the time this project was completed, the U.S. Environmental Protection Agency estimated that computers accounted for 5 per cent of the electrical power consumed by American companies, and that this figure could go as high as 10 per cent if new regulations weren't put in place. Was this the inevitable cost of doing business in the information economy? Not really. Researchers estimated that 40 per cent of computers were left on overnight, wasting valuable energy. These figures showed that if we could encourage people to turn off their computers when not in use, then a substantial impact could be made, not only financially for businesses, but ecologically as well.

The three students adopted a Human-tech approach by trying to understand why many people don't turn off their computers. One reason, they discovered, is that people are impatient: they don't want to have to wait while their computer starts up, every time they use it. If they leave it on all the time it's instantly available to them, a reason also frequently cited as an explanation for leaving computers on over lunch. Another reason is plain forgetfulness – in their rush to leave work, some people simply forget to turn off their computers. Other people leave their computers on all the time because they're under the impression that it's bad for the computer to turn it off and on (a relatively widespread belief despite research showing that if you never turn a computer off the chances are it will last an average of about 2.3 years, but if you consistently turn it off overnight and on weekends, it will last an average of about 9.6 years – an improvement of over 300 per cent). Finally, some people leave their computers on because they think screen savers conserve energy when the computer is on but not in use. This belief is also false; the function of screen saver programs isn't to conserve energy, it's to keep the monitor picture tube from being damaged. By delving into the reasons why so many people *don't* turn off their computers, Kuk et al. uncovered a number of misconceptions, which provided a solid basis for developing a design solution that would be more likely to promote energy-conserving behaviour.

To achieve this goal, they proposed the "Power Pig" – a computer display shown in figure 1 that equates wasting electricity with overeating. It uses a behaviour-shaping constraint: their strategy was to leverage the North American cultural tendency to associate obesity with unattractiveness – an idea based on solid scientific research in social psychology – to convince people that consuming less energy is akin to being trim and slim: if people thought of energy conservation as attractive and desirable, they would be more likely to engage in positive behaviour.

The top part of their display (at the top of figure 1) is a cartoon drawing of a pig. A number of different images are available, from a fat ugly pig representing an energy glutton, to a slim, muscular, attractive pig representing an energy conserver. This portion of the display would be updated frequently to provide people with prompt, salient feedback as to how well they're achieving the desired goal. The students tailored the feedback to the individual, because psychological research indicates that the single most important factor in changing performance in a company is for each employee to develop a standard of his or her own. That way, people have a personal stake and will be motivated by their own wish to succeed.

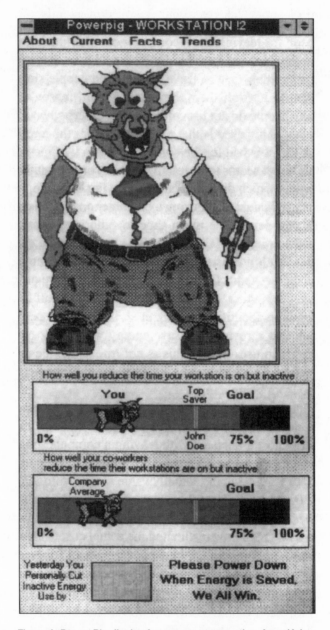

Figure 1. Power Pig display for energy conservation, from Kuk et al. (1994). See text for description.

To keep them challenged, Power Pig doesn't let them see the more attractive pig images until they change their conservation behaviour. You can't sneak a look ahead; you have to earn it.

The second part of the screen, just below the pig cartoon, shows a horizontal bar display with a pig icon. This display is based on the metaphor of a race between the computer user and the top energy conserver in the entire company – John Doe. John Doe serves as a role model, showing that better performance is achievable, to deflect the common rationalization that people use to avoid changing their behaviour: "Why should I change? Nobody else is. I'm doing the same thing as everyone else."

The third part of the Power Pig screen, near the bottom of figure 1, shows a horizontal bar chart displaying the company's average level of performance with respect to its predefined goal. The direction of the pig icon represents the daily trend in energy efficiency: a pig pointing to the right indicates an improvement, while a pig pointing to the left indicates a decrement. This bar display is aligned, and on a common scale, with the one described above, making it easy for the employee to see at a glance whether their behaviour has placed them above or below the company average. The company-wide information in this display is important because it shows everyone that changes in their own behaviour can have an impact on the performance of the whole. This way, people can't resort to another commonly used excuse to keep from changing their behaviour: "Why should I change my behaviour? One person won't make any difference." At the same time, energy gluttons may be embarrassed into changing by seeing just how much more energy they use than others in their company.

Finally, at the bottom of figure 1, a digital display shows the person's own daily efficiency level, providing a clear indication of daily changes in a precise, quantitative form, complementing the graphic displays above it. You can think of this display as the energy conservation "grade" that the person has achieved.

The Power Pig creates a more environmentally friendly computer system by embodying two Human-tech design principles – behaviour-shaping constraints and prompt, salient feedback – that have been proven to result in more effective products. And the design leverages what we know about human psychology, resulting in a "green" version of our benevolent "invisible hand" – a helpful aid that supports us in changing our behaviour – changes that are good for the environment, and thus, for humankind.

WHAT YOU SEE IS WHAT YOU GET:
SAVING PAPER WITH PHOTOCOPIERS

Another group of students set out to tackle another important global problem, the wastage of paper.[3] Studies have shown that an average of over 380 kilograms of paper and paper products are thrown out, per person, per year in the United States alone, which not only destroys our natural resources but also creates pollution. This wastage doesn't all come from one source, but one big culprit is certainly the modern photocopier. We've all made copies that we don't want and wound up throwing paper away. In some cases, mistakes are made because we can't figure out how the page should be oriented: on some copiers, we're supposed to lay down the paper to be copied in a vertical orientation, but with other copiers, we're supposed to lay down the paper in a horizontal orientation. Sometimes face up; sometimes face down. How do we figure out which is the proper orientation? Of course, we can read the manual; it has all of the instructions – and in several languages – but frequently the manual isn't readily available, and even when it is, we usually ignore it and look at the machine for instruction. There are some coloured lines and icons on most photocopiers that are supposed to tell us something, but they aren't always informative, especially if they're on a copier that we've never used before – at a library for instance. So what do we do? Well, we take a guess, press the big green copy button, and hope for the best. Sometimes we guess right and get a good copy, sometimes we guess wrong and we get a copy that has only part of the original and the rest of the page blank.

And then there's the business of enlarging and reducing. Say you have a big newspaper page, but you only want to copy a small article on that page. Which magnification or reduction factor do you pick? 92 per cent? 110 per cent? 79 per cent? After a few tries, you get what you want (unless by then the machine has run out of paper and you have to go find the person who restocks it) – make sure to write down that magical magnification/reduction value or else you'll have to start all over. Do you still have enough change to pay for the copies you need to make?

Don't you hate the feeling of having spoiled all those perfectly good sheets of paper, just because you weren't sure what the photocopy was going to look like before you pressed that green button? It's almost as if there's a ghost in the machine, intent on throwing paper into the garbage bin periodically. Mind you, there's nothing wrong with the mechanical and electronic components inside the copier; they do their job well. But somehow we have to dance uncertainly to get the machine to do what we want – a classic sign that we're dealing with Mechanistic technology that's got a bad fit with people and is slowly contributing to, rather than reducing, environmental degradation. The amount of waste photocopiers generate isn't inevitable; it's a byproduct of the designers' not having paid enough attention to our human needs and capabilities when they created the machine. Think about the millions of photocopiers used in organizations – offices, libraries, schools, universities and commercial copy shops – and you'll start to get a feel for the magnitude of the problem. Given the tremendous volume, even a very small improvement in the percentage of paper wasted can lead to sizable gains that would benefit the environment.

Chris Long, Deryck Ramsahai, Zoher Bhujwalla and Julian Wright adopted a Human-tech approach to tackle the problem. They began by interviewing photocopier users, and by analyzing the tasks involved in using a photocopier. They learned that lack of prompt, salient feedback – the same problem that made it difficult to get a seal on the old-style zip-lock bags – was the leading cause of paper wastage. People were working in the dark – they couldn't see what a copy would look like before they printed it out. As one person interviewed observed: "After a while you learn, but that's twenty papers later and twenty papers wasted."

The students proposed a display built into the photocopier, driven by a scanning device, that would show people what the photocopy would look like *before* it was printed. If it looks crooked, you adjust the original until it looks straight. If you can see that you put the original in sideways instead of vertically, you just realign it. If the contrast isn't quite right, you adjust that. If you don't get your newspaper article coming out at the size you want, no problem; just readjust the magnification/reduction rate until it looks right. And the big environmentally responsible payoff is that you can do *all* of this without wasting a single piece of paper.

I've never seen a photocopier designed with such a display, but perhaps it exists. Or maybe it just doesn't exist yet, because the design isn't technically feasible for some obscure reason, or isn't deemed cost-effective using current technology. Regardless, the students' brilliant idea still shows how creative application of Human-tech thinking at the psychological level can potentially address environmental problems in remarkable ways. I'd be willing to bet my paycheque that such feedback on photocopiers would lead to an enormous reduction in paper wastage, to the benefit of the forests.

MELTDOWN, AMERICAN STYLE:
NUCLEAR POWER PLANTS

The threat to our quality of life is absolute when it comes to our handling of nuclear power. The devastation caused by Chernobyl around the world was actually prefigured nine years earlier in the United States.

In 1977, the industry-funded Electric Power Research Institute published a technical report evaluating how effectively people operated in nuclear power plant control rooms of the day.[4] The sixteen-month study was conducted by three human factors engineers from the Lockheed Missiles and Space Company – Joseph Seminara, Wayne Gonzalez and Stuart Parsons – all of them experienced in the aerospace and military industries. Their goal was to use generally accepted human factors design standards from those industries to evaluate practices in the nuclear industry.

The photograph on the next page shows a nuclear power plant control room. At first blush, it seems complex and intimidating, but it's even more complicated than it looks because the figure doesn't show the entire control room. The barely visible rectangular tiles in the background along the top are alarms that light up when there's something wrong. Below the tiles, there are hundreds of analog gauges that display the state of many plant variables (e.g., temperatures, pressures, flow rates, levels, etc.), showing the operators what's going on out in the plant. The flatter portion at the bottom of the panels houses the knobs, dials, switches and buttons used by operators to change the state of components outside of the control room (e.g., opening and closing valves).

The control room at the Three Mile Island nuclear power plant .[5]

After studying five control rooms of this type, the Lockheed team was deeply disturbed by what it saw. They documented hundreds of examples in their lengthy report, showing that the technology was unnecessarily making operators' jobs *more* difficult, and thereby increasing the chances of human error. Here are some of the more obvious design blunders that were cited, to give you an idea of how ludicrous the conditions were.

- the operators' desks were sometimes placed so that they had their backs toward the gauges they were supposed to be looking at.
- two plants had inadequate heating and air conditioning systems, causing diesel generator fumes to sometimes go into the control room.
- in some cases, operators had to look at gauges that were placed twelve feet above floor level.
- alarms, which were supposed to indicate an abnormal or

emergency situation, were always on, even under normal circumstances, creating a "cry-wolf" situation.

- knobs controlling unrelated pieces of equipment looked exactly the same and were placed right next to each other, making it likely that operators would mistake one for the other.
- knobs controlling related pieces of equipment were placed far away from each other, increasing the difficulty of the operators' job.
- controls were sometimes put in a location and orientation where it would be easy to inadvertently activate them – just by walking by.
- paper tags that were put on gauges to indicate that they were out of order sometimes obscured nearby gauges, making it easy for operators to miss an important indication.

Typically, the technological details were well taken care of, but the human factor was virtually ignored. In fact, several of the control room deficiencies in the nuclear industry identified by the Lockheed team in the late 1970s were almost identical to the cockpit defects observed by Paul Fitts and Richard Jones in aviation in the early 1940s. Nuclear power was thirty-five years behind the times, a slave to the Mechanistic view. From a Human-tech perspective, these conditions were an accident waiting to happen. If there was anything surprising about this situation it was that operators were able to function at all under these appalling circumstances – a testament to human adaptability.[6]

According to Stu Parsons, one of the authors of this seminal report, part of the problem was that most control room instrumentation engineers had never seen an operating control room or talked to operators about their problems – management wouldn't authorize trips of this kind. So there was no way they could have followed a Human-tech design philosophy even if they had wanted to. But the problems ran far deeper than that – the major nuclear power companies such as General Electric,

Westinghouse, and Combustion Engineering had up-to-date human factors engineering expertise in their military product divisions, but little or no human factors competency in their control room design divisions. Also, large architect and engineering firms such as Bechtel, Stone and Webster or Black & Veetch had no human factors engineers on their staffs. They were dominated by classically trained Wizards. Even worse, the U.S. Nuclear Regulatory Commission – the federal government body responsible for regulating the American nuclear industry – openly admitted that it was only interested in reviewing the inside circuitry of the panels from a Mechanistic perspective, not the front of the control panels from a human perspective. No wonder nuclear power was so far behind the times.

You might think that the human factors horror stories collected by the Lockheed team would be enough to convince the Wizards of the limitations of their ways – diesel fumes in the control room was a particularly compelling design defect. Surely, the Electric Power Research Institute report would serve as a wake-up call for the entire nuclear industry to make the transition to Human-tech thinking and immediately improve its control-room designs?

Well, it didn't. Although funding for a follow-on project was provided, the industry didn't respond urgently to the indictment in the report – even though human life and the natural ecology were at grave risk. It would have to undergo a tumultuous conceptual revolution before it learnt to read the writing on the wall penned by the Lockheed team. As we all know, revolutions – conceptual or otherwise – can be painful.

This one was no exception. The pain began at the Three Mile Island nuclear power plant in Harrisburg, Pennsylvania, at 4:00:37 A.M. on March 28, 1979, unleashing a wave of fear through American society that continues to ripple to this day.

To begin with, a few financial statistics may help give a rough idea of the magnitude of the destruction that was experienced.[7]

The Three Mile Island plant cost US $700 million to build in the 1970s, but it functioned at full power for only four months before the accident occurred. The reactor core was so badly damaged during the accident that a decision was made never to operate the plant again. It took about fifteen years and US $973 million to perform all the steps required to clean up the contamination. As I write, the plant hasn't been decommissioned yet, but the cost of doing so has been estimated to be $398 million (in 1996 US dollars). All told, this was a multibillion-dollar loss – a high price to pay for a mere four months of electricity production.

What could possibly have happened inside that nuclear reactor core to cause such a calamity? The full details may never be known with certainty, but ten years after the accident enough evidence had been gathered for experts to determine the internal damage that was caused to the reactor.[8] Scientists were able to inspect different areas of the core visually, using remotely controlled closed-circuit television cameras, and they were able to take material samples from the core and analyze their physical composition. These painstaking investigations painted a horrifying picture of the physical devastation that must have been unleashed inside the reactor core during the accident – nuclear technology run amok, releasing some of the strongest forces in Nature's arsenal.

For starters, at least 45 per cent of the nuclear reactor core had melted. How much material does that represent? The answer to this question provides a clearer indication of the gigantic proportions of the devastation – sixty-two *tonnes* of material were melted. The image created by that statistic takes on an even more sinister significance when we consider that the metals and ceramics that melted were among the sturdiest materials known, designed to withstand the hottest conditions imaginable – yet the forces that were let loose inside the reactor core were too much for them. The damage was so bad that twenty tonnes of core material melted and eventually slithered all the way down the reactor vessel, settling into a molten cesspool at the bottom. Inside the core, scientists found samples of molten ceramic compounds consisting of uranium, zirconium and oxygen, indicating that some regions of the upper reactor core reached infernal peak temperatures of over 2,800°C.

To understand how this could have happened, it may help to review the way water boils as a function of temperature and pressure. As we all know, water boils at 100°C at atmospheric pressure, sea level. However, if pressure is lowered, then the boiling point is also lowered. People who live at a high altitude, where the air is thinner, know this; because the pressure is lower than at sea level, water boils at a lower temperature at high altitude. You don't have to get the water to 100°C to make a cup of coffee. The relationship between temperature, pressure and the boiling point of water works in both directions. As pressure goes up, the boiling point increases accordingly. This relationship allows very hot water (well above 100°C) to remain in a liquid state.

The operation of a nuclear reactor is designed to take advantage of this relationship. The reactor pressure is kept at a very high level, so that the water can be very hot – several hundred degrees above its usual boiling point – yet still remain in a liquid state. You don't need to do that for your morning cup of coffee, but to produce electricity safely in a nuclear power plant, it's absolutely crucial to maintain a cover of water in a liquid state over the reactor, to keep it in control. If the boiling point is reached, and the liquid water gets transformed into steam, the core will no longer be covered by the liquid water. At this point, the reactor core will start to heat up. Unless liquid water is added, the temperature will continue to soar and eventually cause the mighty materials that make up the core to melt.

The lesson is very simple – don't let the water boil in a nuclear reactor. Otherwise, you will be faced with a modern-day technological equivalent of Dante's Inferno, which may wreak destruction that can threaten not only public safety but also the natural environment.

Yet that's exactly what happened on that frightening day in March 1979 at Three Mile Island. The control room operators knew that water should not be allowed to boil in the reactor core, but they didn't figure out that the reactor had been boiling until 142 minutes after the accident had started – a very long time in the nuclear business. A number of pundits attributed this delayed diagnosis to the catch-all explanation "human error." But from a Human-tech perspective, we get a more nuanced and clearer picture of who and what was to blame by considering the context the operators were working in, and how they experienced the unfolding of events that contributed to their so-called " human error."[9]

Before the accident began, the reactor was operating at close to full power, and there didn't appear to be any cause for concern. But unbeknownst to the operators, a valve that was supposed to be open had been left closed by a maintenance worker who was no longer on shift. And it was four o'clock in the morning, the lowest point in the human circadian cycle. No matter how well-intentioned they are, how attentive they try to be, how much coffee they've had, people aren't at their best at 4 A.M. – not even people whose job it is to control a nuclear power plant. After the complex accident sequence began, more than a hundred alarms went off in the control room. There was no way for the operators to turn off the less important alarms. The effects of the accident, as it began, were so severe that several of the instruments were off-scale. Fortunately, however, a computer printer was recording the events, so operators had a paper copy that they could use to interpret changes after they occurred. However, that printer was running more than *two hours* behind the pace of the events!

What about those huge computerized control room consoles? Didn't they give the operators clear, unambiguous readings of the state the plant was in? Well, judge for yourself, bearing in mind that as an operator, you're responsible for one of the most complex technological systems ever invented: one critical indicator was covered up by a paper tag hanging from another instrument on the control panels; a second important indicator was located on a panel *behind* the seven-foot-high control panels that operators usually look at, where all of the critical displays are; a third crucial indicator misled the operators because it was designed to show what should have been going on, not what was really going on. Given such conditions, is it any wonder that the operators didn't realize right away that the reactor core had started to boil? You'd have to be half-blind to attribute the accident to "human error."

Leo Beltracchi, a now-retired engineer, took that lesson to heart. Before retiring, Beltracchi had worked for the United States Nuclear Regulatory Commission (USNRC), and served on the USNRC "Lessons Learned" Task Force that was struck after the Three Mile Island accident. He was particularly concerned by the notion that "human error" was to blame. He knew a great deal about the devastating effect that out-of-control technology can have on nuclear safety: he had spent dozens of hours sitting in nuclear power plant control rooms, talking to operators, watching them do their job and acquiring a great deal of real-world experience. And unlike most of us who studied thermodynamics as undergraduate engineering majors, Beltracchi actually understood and remembered some of what he had learned. Relying on his scientific understanding and his appreciation for the human perspective, he set out to try to understand why the power plant operators had failed to cope with the unfamiliar and unexpected situations that had occurred at Three Mile Island, and to try to prevent it ever happening again.[10]

To understand Beltracchi's thinking, we need to understand what engineers refer to as the *saturation properties* of water. These properties describe the relationships between pressure, temperature and boiling point that I referred to earlier. Most people find these relationships to be counter-intuitive. For example, operators could observe that the pressure is at 7 MegaPascals (Mpa) and that the temperature is 269°C. Does this mean that the water is in a liquid, steam, or combined liquid-plus-steam state? We need help in understanding and using these relationships, so engineers have created what are called steam tables. A full set of steam tables fills several pages, which I will spare you here, each filled with columns and columns of numbers representing weird-sounding thermo-dynamic concepts, such as enthalpy, entropy and specific volume. A simpli-fied version is shown in table 1. The first column shows the pressure and the second column shows the satura-tion temperature, which is the boiling point of water at the corresponding pressure. Using these data, you can tell if water is in a liquid or a steam state.

PRESSURE Mpa (g)	T_{sat} °C
9.00	301
8.80	300
8.60	298
8.40	296
8.20	294
8.00	293
7.80	291
7.60	289
7.40	287
7.20	286
7.00	284
6.90	283
6.80	282
6.70	281
6.60	280
6.50	279
6.40	278
6.30	277
6.20	276
6.10	275
6.00	274

Table 1. An example of a steam table showing the saturation properties of water — the boiling point of water at the correspon-ding pressure.

As you can imagine, this informa-tion is of critical importance in a nuclear power plant – given that it's essential that the water in the reactor *not* reach boiling point. So how would you, as an operator, use a steam table to determine whether the plant has reached, or is approaching, this dangerous state? The details of the procedure will differ somewhat for different control rooms, but the basic steps are as follows. You have to:

1 Remember where to find the gauge that represents the pressure in the reactor.
2 Walk up to the gauge and read the current pressure value (e.g., 6.00 Mpa).
3 Memorize or record that value because it will be used later.
4 Remember where to find the gauge that represents the temperature in the reactor.
5 If the temperature gauge is far away from the pressure gauge, walk to its location.
6 Read the current value (e.g., 250°C).
7 Memorize or record this value as well because it too will be used later.
8 With the pressure and temperature values in mind, find a steam table like table 1.
9 Look down the first column to find the row that corresponds to the pressure reading that you memorized or recorded in step 3 (e.g., 6.00 Mpa).
10 Look up the saturation temperature for that pressure value (i.e., 274°C).
11 Compare the saturation temperature with the temperature reading that you memorized or recorded in step 7. If the observed temperature is lower than the saturation temperature (i.e., 250°C < 274°C), then the water is in a liquid state. If the observed temperature is the same as the saturation temperature, then the water is in a liquid state, but it will start boiling if any more energy is added. If the observed temperature is greater than the saturation temperature, then the water has already started to boil and there's a mixture of liquid and steam. You are in deep trouble.

The procedure for evaluating reactor safety using a steam table clearly requires quite a few steps. Operators have to memorize or record numbers for later use. They may have to walk around to different locations in the control room. They have to look up values in a numerical table where, at a glance, each row looks like every other row. They have to do some unaided mental arithmetic, an error-prone process. In short, the psychological demands associated with using steam tables are not trivial.

But the burden doesn't end there. The only time that operators are required to perform all these steps is during emergency situations when alarms are blaring and the terrifying threat to the plant's safety, and thus to the public safety and the natural ecology, is at its peak. It wouldn't be surprising if operators occasionally made a mistake when performing this cumbersome procedure under such demanding conditions. Nevertheless, the control room design at the Three Mile Island power-plant required operators to do exactly that.

Leo Beltracchi thought there just had to be a better way to present thermodynamic information in a way that operators could easily interpret. He remembered taking a course in thermodynamics in 1951 as a mechanical engineering undergraduate at Clarkson University in Pottsdam, New York. The course was taught by Professor Edward McHugh, a brilliant, eccentric man who had a gift for making complex subjects simple and easy to understand. McHugh made quite an impression on his students, not only because of the clarity of his lectures but also because of his theatrical style. He would come into class and sit on the table with his legs folded yoga style in the lotus position. Then the lecture would begin. It was in McHugh's class that Beltracchi learned the fundamentals of thermodynamics, and this knowledge was to stick with him for the rest of his life.

Beltracchi's job responsibilities at the USNRC had not included designing better control rooms, but he had both the dedication and the will to develop new design ideas on his own time, and that perseverance would stand him in good stead when he focused on Three Mile Island. His friends sometimes affectionately refer to him as a "weekend researcher." And every time he presented a paper at an international conference or published a paper in a scientific journal, Beltracchi was required to add a disclaimer: "The opinions and viewpoints presented herein are the author's and should not necessarily be interpreted as the criteria and guidelines of the USNRC." And so it was on his own initiative, after he had served on the Lessons Learned Task Force, that he came up with a revolutionary design idea for a computer-based display that could help operators quickly understand the psychologically counter-intuitive saturation properties of water.

Beltracchi's creation is a wonderful example of how a good fit can be achieved by adapting design to human capabilities: in this case one of the most powerful and efficient abilities nature has equipped us with: pattern recognition. His creative computer display is based on three simple ideas: a) it's important to convey the information that's contained in steam tables; b) steam tables are cumbersome for people to use; c) the same information can be presented in a quickly comprehended graphic format, so people see and immediately understand what's going on in the plant. Beltracchi's display employs a diagram that has been used for decades in textbooks to teach thermodynamics to engineers.

That diagram is illustrated in figure 2 on the next page, which shows the saturation curve for water in a two-dimensional graph defined by temperature (T) and entropy (s) axes (s is the standard symbol for entropy in thermodynamics). Using this graph, it's very easy to tell what state water is in: if we're on the left of the curve, the water is liquid; if we're under the curve, then there's a mixture of liquid and steam; and if we're on the right of the curve, the water is steam.

Beltracchi's brilliant insight was that the diagram could be brought to life to create a dynamic computer-based display. The complete design requires a deeper understanding of thermodynamics than I've covered here, so I'll only present a simplified account of the basic idea. Sensors are used to measure the

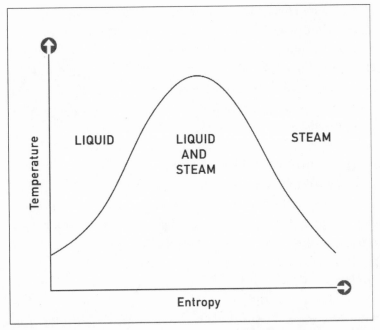

LIQUID

LIQUID
AND
STEAM

STEAM

Temperature

Entropy

Figure 2. The temperature-entropy diagram showing the saturation properties of water.

temperature and pressure values. These correspond to a point on the graph in figure 2, because pressure can be converted to entropy (which is defined as the quantity of energy no longer available to do work). Thus, the static T–s diagram found in textbooks can be used to create a dynamic display that shows the current state of the reactor, in real time, against a graphic that shows the saturation curve as a constant backdrop. As the temperature and pressure in the reactor change, the location of the point in the animated display is updated accordingly.

So how would operators use Leo Beltracchi's display to determine if a nuclear power plant is approaching, or has already reached, the perilous state of boiling in the reactor? The procedure is far, far simpler than the eleven-step mental gymnastics routine required by steam tables.

1 Walk up to the computer display.
2 If the data point is under the curve, then there is a mixture of liquid and steam, and safety is being threatened. If the point is on the left edge of the curve, then the water is in a liquid state, but it will start to boil if more energy is added. If the point representing the current state is to the left of the curve, then the water is in a liquid state, as desired.

Two steps instead of eleven. Operators don't need detailed knowledge of the saturation curve and don't need to read scales. All they have to do is recognize graphic patterns. No recording, memorizing or calculations. The design is based on our human talent for pattern recognition, a talent even Homer Simpson has in abundance.

Indeed, the benefits of Beltracchi's display have been confirmed through a scientific experiment that my colleagues and I conducted when I was a Ph.D. student at the University of Illinois. Professional nuclear power plant operators used Beltracchi's design and two other more traditional displays to diagnose safety-critical faults generated on a realistic industry-scale nuclear power plant simulator. Beltracchi's display improved the accuracy of the diagnoses by about 80 per cent.[11]

This innovative display has had a notable impact on the nuclear industry. The design concept has been implemented in advanced control room simulators at the Halden Reactor Project Laboratory in Norway and at the Toshiba Nuclear Energy Laboratory in Japan, and in an operating reactor at the Experimental Breeder Reactor II in Idaho Falls in the United States. In the future, it wouldn't be surprising to find this display being used at every nuclear power plant control room in the world – and it should be.

After the devastating transitional instability caused by the Three Mile Island accident, the nuclear industry experienced a Human-tech Revolution. Leo Beltracchi's innovation is now just one of many that have been introduced to improve safety by incorporating human psychology, thereby helping operators do their job more effectively and improving plant safety. Other designs include data displays that are easier to read, emergency procedures that are more clearly written, more effective training programs to teach operators about the principles governing the behaviour of the plant, and better alarm systems that make abnormal situations easier to detect and diagnose. The nuclear industry is no longer dominated by a strictly Mechanistic view. Together, these design changes have had a significant impact in reducing what bad fit there was between people and technology.

These improvements are more relevant than ever. Despite the growing anti-nuclear sentiment after Three Mile Island and Chernobyl, in 2002 there were still 432 nuclear power plants in operation worldwide, supplying a total of 340,347 megawatts of power – about 17 per cent of the world's electricity.[12] The U.S. was the leader with 109 reactors, followed by Japan with 59 and France with 56. All these plants are carefully designed to produce electricity using a nuclear reaction under controlled conditions. Nevertheless, unless we're able to handle the complexity of our own creations, the type of widespread havoc unleashed by Chernobyl could haunt us again at another time and in another place, perhaps even in our own backyards. It's essential for designers to continue to pay attention to the human factor, so that fatal catastrophes that threaten both human life and the environment are less likely to occur. And since, as we now know, a nuclear accident anywhere can be a nuclear accident everywhere, all of us will benefit.

SPOT THE TERRORIST: AIRPORT SECURITY

On the morning of September 11, 2001, the gaping holes in American airport security systems were visible for all the world to see. This we know. What is still not clear, however, is *why* security was so woefully inadequate. And without an accurate diagnosis, we risk investing a tremendous amount of money in massive changes that don't fix the problem. Pinpointing the nature of the underlying deficiencies is critical, and could potentially avert future terrorist acts and save untold numbers of lives. As I mentioned earlier, changes are being made to airport security systems, so it's difficult to evaluate their current state, but the situation before September 11th is well known.[13]

Let's start with job motivation. Baggage inspectors were paid minimum wage; for them an entry-level job at McDonald's would be a step up the career ladder. In 2001, the average wage in the industry was still only a paltry US $6.75 per hour, and in some airports, the wage was as low as US $5.15 per hour. As a result, the job turnover rate was incredibly high; employees stayed on the job only until better career opportunities arose, which given the meagre pay didn't take very long. In 2000, the turnover rate at all major U.S. airports exceeded 100 per cent. One major airport had a turnover rate of 400 per cent.

So, job motivation was clearly very low; which leads to another problem – inadequate experience. For example, about 90 per cent of the security personnel at one checkpoint were found to have less than six months' work experience. And to make matters even worse, airport security personnel received only twelve hours of classroom training.

This abysmal level of proper experience and training persisted because the procedures that had been developed (before 9–11) by the Federal Aviation Administration – the federal government regulator for aviation in the United States – to test baggage inspectors were, well, laughable. Once in a while, an FAA employee would come by, disguised as a passenger, and attempt to get through the security checkpoint with one illegal object in an otherwise completely empty suitcase. That's right, just *one* object – no clothes, no camera, no hair dryer, no other objects at all that could distract the inspectors and make their job more difficult (and realistic). But that's not all. The test object was one of eight standardized items sometimes referred to as the FAA-8. One of those objects – I kid you not – was a bell alarm clock with a few sticks of dynamite tied to it, much like the kind Wile E. Coyote ran around with in those old Bugs Bunny–Road Runner cartoons. This procedure was used to make sure that the screener's bag-checking skills were up to par. At least the dynamite wasn't wrapped in a sticker labelled "ACME Company."

Poor job training and inadequate experience weren't the only sources for the problems. The psychological demands imposed by the security inspection job are also a crucial factor. First, there's the sheer level of traffic. Each day, over 1.5 million passengers have to be screened at airports across the U.S. As a result, an inspector at a major airport might have to examine more than 10,000 bags in a month. Then, there's the incredible time pressure. Both passengers and crew get angry and abusive if they have to wait in line while security inspectors check a bag manually or order someone through the scanner again, holding everyone up, making the line even longer, and making the passengers even more anxious and even more irritable. Passengers – and we've probably all seen it – will sometimes yell at the security personnel, telling the screeners to let them through right away, otherwise they'll miss their flight. Before September 11th, most people were worried about catching a flight that left in two minutes, not about thwarting potential terrorists.

Lack of immediate, salient feedback was also a problem. Except for the ridiculous FAA-8 test I described above, security personnel received very little information about the accuracy of their judgements on the job. Nobody knew how many illegal items each inspector let go by undetected. Yet one of the things that we know about human psychology is that it's critical for people to receive timely feedback on their job performance. Doug Harris, a human factors engineer who has studied airport security in great detail, came up with a vivid and compelling analogy to illustrate the psychological consequences of lack of feedback: "Consider . . . how little people would improve their bowling performance and how soon they would stop bowling altogether if there were no system of keeping score and no feedback on how many and which pins were knocked down with each roll of the ball."[14]

All these psychological factors – poor motivation, deficient training, insufficient experience, high traffic, intense time pressure and paltry job feedback – were significant contributors to the appalling level of airport security, but perhaps the single most important threat to effective performance is the simple fact that illegal objects are exceedingly rare. Psychologically, this makes the task of baggage inspection very demanding because people have to remain vigilant for long, uninterrupted periods during which they're looking solely at bags that have only legal objects in them. Scientific research conducted since the late 1940s had shown that people are very poor at performing vigilance tasks; after half an hour, performance drops by almost 50 per cent. The phenomenon has been heavily investigated and is known as "vigilance decrement."

The bad fit between people and the airport security system in which they worked gives a clear indication of why it was in such sad shape before September 11th; the overall system was designed in such a way that it violated almost every relevant known fact about human psychology. But the great advantage of the Human-tech perspective is that it doesn't just allow us to understand the causes of problems, it also provides systematic principles for solving them too.

Signal detection theory (SDT) is a Human-tech technique that's ideally suited to improving human performance on vigilance tasks like baggage screening.[15] SDT tells us that if we're going to give inspectors feedback about their job performance, or evaluate how well they're performing, we can't just measure the proportion of illegal items that they detected successfully. Inspection tasks like those involved in checking luggage at airports represent an intrinsic trade-off between two conflicting factors: correctly identifying an illegal object and incorrectly identifying a legal object. In SDT, these two factors are referred to as "hits" and "false alarms" respectively. Inspection performance can be meaningfully evaluated only by considering both these factors together, and SDT provides a set of mathematical formulae for doing precisely that.

Most people think that the "hit rate" would be an effective measure of job performance. For example, if you detected all the illegal items that come your way – a 100 per cent hit rate – then you must be doing an outstanding job as an inspector, right? It's precisely this kind of logic that was behind a November 2001 memo from Doug Ose – a Republican member of the U.S. House of Representatives – reporting that the Government Reform Subcommittee on Energy Policy, Natural Resources and Regulatory Affairs was considering "a zero tolerance policy for mistakes" in airport security.[16] From the viewpoint of a layperson, this might sound like a good idea. However, SDT tells us that such a policy wouldn't make any sense at all.

There's one strategy that would guarantee a 100 per cent hit rate, a strategy so simple that even a child could carry it out; all you'd have to do is inspect each and every single piece of luggage that comes your way. Imagine the line-ups that would be generated if airport security inspectors did that. But what choice would they have if their jobs depended on maintaining a 100 per cent hit rate? The problem with this strategy, of course, is that it would generate a huge false alarm rate as well as a perfect hit rate, a fact that can be predicted with mathematical precision using SDT. If Doug Ose's memo is any indication, most people still don't have a clue about this relationship – yet this is the knowledge we need to put to use if we seriously want to improve the level of airport security.

The knowledge *is* available – and was available pre-September 11th. In addition to clarifying how to measure inspection performance, SDT can help us reduce the deficit of alertness and attention induced by the sheer rarity of illegal objects. Scientific research on other vigilance tasks has shown that people remain more attentive and performance improves if "false signals" are introduced periodically to stimulate people to remain alert. In the case of airport security, images of illegal objects superimposed or blended in with legal objects could occasionally be projected on the monitoring screen. These recurrent false signals would serve to "wake up" the inspectors, and help ensure that they don't fall prey to vigilance decrement. The basic idea is described quite simply in a technical report written by Klein Associates, a consulting company owned by Gary Klein, author of *Sources of Power: How People Make Decisions:*

> As the operator screens parcels, the system selects a suitable parcel that passes through at a predetermined time and injects the image of a threat item. If the operator detects a threat s/he presses a specific button on the X-ray machine's control panel. The system then displays a message informing the operator whether s/he is being tested. It then asks the operator to locate the threat in the parcel and to classify it as a gun, knife, explosive, or miscellaneous weapon. If the threat is a false image . . . it disappears. If it is the image of a real threat item, then it does not disappear and the operator is warned that a test is not in progress.[17]

This design has other benefits, including periodically providing some badly needed on-the-job training and it could be used to create a database for the evaluation of each inspector's performance, not under sterile lab conditions, but under realistic conditions with dozens of irate and anxious passengers waiting to get through the security checkpoint and onto their flight.

In fact, the FAA did develop such a Threat Image Projection (TIP) system in cooperation with human factors engineers – a design that made so much sense that it received the *Aviation Week* Technology Innovation Award in 2000.[18] A total of 1,200 TIP systems needed to be deployed to equip all the necessary checkpoints in the United States, but unfortunately the introduction of the award-winning system was "slower than projected." Improving the people-technology fit was apparently not a high priority in the airport security business before September 11th, 2001.

As with the other situations that I've described, the psychological roadblocks imposed by Mechanistic thinking were to blame. Many designers of airport security systems were bent on developing "clever" technology, like automated baggage-checking systems based on artificial intelligence techniques, in the hope of completely eliminating human involvement – a goal that the best and brightest of the Wizards are still nowhere near achieving, despite tremendous financial investment and years of technical efforts. Even the FAA, the government regulator, was unsuccessful in overhauling the ineffective and unsafe airport security systems, in part because upper management didn't appreciate that the biggest problem wasn't a lack of technical cleverness but rather a poor fit between people and technology. Like aviation during World War II, the airport security business was stuck in the Mechanistic world view.

Unfortunately, however, Mechanistic thinking isn't the only obstacle standing in the way of designing sound, complex safety-critical systems. And the area of greatest concern to most of us is hospital safety. It is worth repeating the astounding statistic I mentioned in chapter 1: the preventable mortality rate in U.S. hospitals today is the same statistically as a wide-body jet aircraft crashing every day or two, with no survivors.

THE PAIN PUMP: DEATH BY MORPHINE OVERDOSE

At 2:34 A.M. on February 27, 2000, nineteen-year-old Danielle McCray was admitted to Tallahassee Memorial Hospital to have her baby.[19] An uneventful C-section was performed at approximately 4:30 P.M. and a healthy baby girl was delivered. At about 6:45 P.M., Danielle complained of pain, so fifteen minutes later she was connected to a patient-controlled analgesia (PCA) infusion pump. This device, once programmed by a nurse, allows patients to self-administer painkiller – in this case, morphine. At 8:30 P.M., Danielle was awake, alert, and feeding her newborn daughter for the first time. About six hours later, she was pronounced dead after a resuscitation effort that lasted approximately thirty-five minutes. The autopsy and toxicology analysis results showed that she died of an overdose; she had a total concentration of 761 ng/mL of morphine in her bloodstream – almost four times the lethal amount. One of the primary contributing factors was human error in programming the PCA device.[20]

Danielle McCray wasn't the first person reported to have died while connected to this model of PCA pump. On the contrary, several years before her death, the medical literature documented a number of cases where patients reportedly lost their lives from morphine overdoses after errors had been made while programming the same device.[21] In fact, in 1997, three years before Danielle McCray's death, the Emergency Care Research Institute (ECRI) – a world leader in medical device safety – had issued a medical device alert, stating that this particular PCA device was "susceptible to misprogramming" and that "the likelihood of this sort of misprogramming is increased by the fact that the user interface and programming logic of the pump are particularly complex and tedious. We believe that the likelihood of user error is increased by the repetitive and time-consuming programming process required by this pump."[22] More recently, in 2001, the ECRI evaluated nine different PCA pumps from six different manufacturers, and rated this particular design as only conditionally acceptable, stating that "This pump has a significant safety problem."[23] But even though six of the other eight pumps received a higher overall rating, and only two were rated lower, this one had achieved the greatest market penetration: according to the manufacturer, by 2000, the device was in use around the world, including in nearly four thousand hospitals in the U.S. alone, and accounted for about 75 per cent of all PCA use there.[24] In late 2002, ECRI was still receiving reports of deaths from programming errors on this PCA pump.[25] The exact number of deaths will likely never be known, but the total number of people who may have died from programming errors on this device between 1988 (when it was first introduced) and July 2000 has been estimated scientifically at between 65 and 667 (not including other types of errors that can also lead to death).[26]

The paper trail of warning signs goes back even farther, however. In 1996 – four years before the death of Danielle McCray – the Institute for Safe Medication Practices wrote that the PCA manufacturer had been told that a bad fit between their device and users could potentially threaten patient safety.[27] No technology is risk free, but in this case, the problem was well known.

And so was a potential solution to the problem. In early 1998, Laura Lin, John Doyle, several of my students and I had together published a scientific, peer-reviewed article on this same PCA device[28] At that time, we didn't know of the reported patient deaths that had already been documented in published reports. Our goal was to design medical devices to have an affinity with human psychology by using Human-tech principles that had been developed in, and proven to work, outside health care. We examined the existing design of this PCA device and found that its complexity did indeed make it confusing for nurses to program, so we developed a prototype for an alternative design that aimed to ease the programmer's task.

The end result was a simpler design. Whereas the commercially available design could require as many as twenty-seven programming steps, our new design required a maximum of twelve steps, a reduction of 56 per cent. This certainly seemed like a significant improvement, and the controlled experimental test results showed that the new, redesigned device led to fewer errors, faster task completion times, lower mental workload and an overall better fit with the nurses who used it.

These scientific findings are particularly important where medical errors are concerned because they show that human frailty is not the only cause. After all, if the problem was the people, then changing the design of the device shouldn't make any difference whatsoever because the same people used the old and the new designs. But that's not what our results indicated. Instead, we found that the redesigned device cut the total number of programming errors by *half*.[29] Even more important, the particular type of programming error that had been linked to reported patient deaths was *eliminated*, suggesting that PCA devices could be made safer if designed with greater attention to Human-tech thinking.

How did the manufacturer, the Food and Drug Administration (the federal regulator of medical devices in the U.S.) and Health Canada (the federal regulator of medical devices in Canada) react to these tragic events and to the availability of this information? Clearly, none of these organizations likes to see anyone harmed and they are all deeply committed to patient safety, but cultural assumptions do affect how each of us looks at the world, and therefore the decisions we make. As far as I know, the company didn't immediately change the design of the commercially available PCA, issue a product recall, or even write a letter of warning (as I will discuss in a later chapter, there are legal reasons that may have kept them from doing so); and the FDA and Health Canada both investigated the issue at length but didn't enforce any such actions. However, after ECRI issued its product alert in 1997, the manufacturer wrote a letter to clinicians implying that the problem wasn't with the design of the device, but with a lack of proper user training.[30] After the sad death of Danielle McCray, the manufacturer told a reporter from the *Tallahassee Democrat* that "the [device] has no design flaws . . . the pump is safe if used as directed."[31] To be sure, reports of fatalities are rare; the manufacturer estimates that the device has been used over twenty-two-million times without incident, putting the likelihood of death from programming errors at 1 in 33,000 to 1 in 338,800.[32] However, many potentially dangerous products are also safe if used precisely as directed, but that shouldn't stop designers from building in a sturdy safety net in case people make mistakes. This particular device does have several important safety features, but none of them prevented Danielle's death. Fortunately, the device is no longer on the market (although it is still in daily use world-wide), and the manufacturer started designing a new and improved device in 1996 – a device that was originally due out in 2001, but that's now expected to be available shortly (and may already be on the market by the time you read this).[33]

There was, however, at least one party who took immediate action after Danielle's death – the Tallahassee state attorney. He and his investigators opened a criminal investigation to see if there was enough evidence to lay charges against the nurse who had cared for her. Despite the fact that several reputable organizations and researchers had drawn attention to the design of the device as a potential threat to patient safety, and despite the existence of scientific evidence suggesting that an improvement in the design of the device could cut total errors in half and perhaps eliminate potentially fatal errors, the investigative spotlight was on the person, not the device.[34]

Unfortunately, this punitive attitude is quite typical in the health-care sector. Most people think that the obstacle to safe health care is a group of "bad apples," incompetent or careless nurses and doctors who make mistakes at the drop of a hat, mistakes that injure or kill their patients. This "bad apple" theory has been sustained by highly publicized cases of medical malpractice and willful negligence – some of them horrific like the recent, macabre case of the English doctor who was unmasked as a mass murderer, having killed as many as three hundred of his patients. Such ghastly stories are fodder for newspaper headlines.

The "bad apple" theory is also appealing because of its simplicity. If the vast majority of medical errors are indeed attributable to poor or lazy care, then the remedy is straightforward, at least in principle. We just have to identify the quacks, take away their licences to practice medicine, and in the more egregious cases put the evildoers behind bars. Then the march toward improved patient safety can proceed. Surely, everyone would agree that punishment and censure are just remedies for acts of willful negligence or incompetence.

However, getting rid of the bad apples, although necessary and important, will do little to improve patient safety. The fact is that incarcerating all the "Angels of Death" tomorrow would barely put a dent in the statistics identified by the Institute of Medicine report I cited earlier. The problem of medical error is much more complex than the "bad apple" theory would have us believe.

The research on patient safety unequivocally shows that the vast majority of cases where patients are injured or killed are attributable to honest mistakes made by good people, not negligent or criminal acts committed by incompetent or homicidal health-care practitioners. Bad apples are actually extremely rare in medicine. So why do between 44,000 and 98,000 people still die annually from *preventable* error in U.S. hospitals?

Lucian Leape, adjunct professor of health policy at the Harvard School of Public Health, probably knows more about this topic than almost any other living person. He was one of the co-authors of the landmark Harvard Medical Practice Study that first uncovered the number of people who die each year from preventable human error in U.S. hospitals. He also served on the panel of the IOM report that brought those terrifying mortality statistics to the attention of the general public in the U.S.

In 1994, Lucian Leape, a doctor himself, published a highly influential article about medical error that explained why implementing what I would call a Human-tech view in health care was such an uphill battle:

> Physicians are expected to function without error, an expectation that physicians translate into the need to be infallible. One result is that physicians . . . come to view an error as a failure of character – you weren't careful enough, you didn't try hard enough. . . . It has been suggested that this need to be infallible creates a strong pressure to intellectual dishonesty, to cover up mistakes rather than to admit them. . . . Lessons learned are shared privately, if at all, and external objective evaluation of what went wrong often does not occur. . . . the perfectibility model: if physicians and nurses could be properly trained and motivated, then they would make no mistakes. The methods used to achieve this goal are training and punishment.[35]

He went on to explain how this culture has made it very difficult, if not impossible, for medical professionals to understand and embrace an approach to improving patient safety that focuses on the person-technology fit. After all, if you're supposed to be perfect, then there's no logical reason to embrace design changes. You should be able to achieve flawless performance, even if the technological system around you is poorly designed.

The manufacturer's reaction to the reported deaths of patients connected to PCA devices appears to illustrate this way of thinking: if there was human error, then more training has to be the answer, presumably because humans must be to blame. The logic is seductively straightforward. In fact, it follows elegantly from what, in chapter 1, I called the "super-Humanistic" approach in medicine that expects people to be flawless, even if systems are designed to be incompatible with everything we know about human nature. When I first encountered this view, I interpreted it as a sign of supreme arrogance: if doctors think they're better than everyone else and aren't willing to admit that they need help in reducing medical error, the result will indeed be thousands of lives lost every year. On reflection, I realized that my attribution of arrogance was misguided, but that the second part of the equation – the reluctance to seek help in reducing medical error – had some foundation. As my late friend Jane Poulson – Canada's first practising blind physician – told me, "We get taught in medical school from day one that we can't make a mistake because otherwise we might kill people."[36] It's difficult to disagree with this line of reasoning. So how have we developed a medical culture that has largely adopted the superhuman expectations identified by Leape?

There are two important considerations to keep in balance when considering this question. In a culture that relies so heavily on technology, there is both the human factor and the technological imperatives. We can't fault medical culture for urging health-care providers to try as hard as they can to avoid errors, but we can and should fault it for leading doctors and nurses to believe that flawless performance is a realistic expectation. Downloading the entire burden of patient safety entirely onto the health-care practitioner causes doctors and nurses to believe that those who can't live up to the standard of constant perfection are flawed individuals, even guilty of moral failings. Our everyday experiences and all the scientific evidence about human nature point to a very different conclusion; good people can sometimes do bad things – it's as simple as that. Indeed, the Tallahassee state attorney's background and criminal-history check on the nurse who cared for Danielle McCray revealed a spotless, exemplary record:

> There are no warrants or investigations and her background was uneventful. . . . She was an honor roll student. The Department of Health, Board of Nursing did not have any complaints or investigations. Her personnel file also did not have any complaints. . . . [She] stated that she has used the PCA Infuser pumps at least fifty plus times per year since 1996. She has never had a problem nor has she heard of a malfunction with the sixty or so PCA's in use at [the hospital].[37]

This is hardly the record of a bad apple.

Admitting to fallibility doesn't mean that doctors and nurses shouldn't continue to be attentive and diligent. It means that, despite their best efforts, mistakes will still be made. But the world of health care doesn't revolve solely around health-care providers; medical errors can also occur because of an *interaction* between people and the complex technological systems in which they work. The expectation of perfection should be replaced by a need to design products and systems that help health-care professionals pursue the goal that they sincerely want to achieve – taking care of patients safely. We must make demands of the technology too and fully recognize its crucial role. Health-care professionals aren't used to thinking this way because their vision is obstructed by the psychological blinders the medical culture has imposed.

As Dr. Lucian Leape noted, the road to progress and change is a clear, but difficult, one to follow: "Physicians and nurses need to accept the notion that error is an inevitable accompaniment of the human condition, even among conscientious professionals with high standards. Errors must be accepted as evidence of system flaws not character flaws. Until and unless that happens, it is unlikely that any substantial progress will be made in reducing medical errors."[38] In other words, unless and until a Human-tech Revolution occurs in health care, the idea of designing systems that recognize the human factor will have a hard time showing up on the radar screen, never mind have a positive impact on patient safety. But modifying an entire profession's basic assumptions about how the world works – like any other conceptual revolution – will take time, patience and dedicated effort.

PUTTING PATIENTS FIRST:
ANAESTHESIOLOGY LEADS THE WAY

Although the health care sector as a whole is generally still in the clutches of the super-Humanistic imperative, there are a few very encouraging signs, particularly from the anaesthesiology community. In just ten years, the chances of dying from anaesthesia went from about one in 10,000 or 20,000 to less than one in 200,000. There's no one single reason for this revolutionary improvement in safety, but increased attention to human psychology has played a prominent role.

Dr. Jeff Cooper, a biomedical engineer, was influential in pioneering this new thinking in anaesthesiology in 1975, decades before the general public became aware of the devastating extent of medical error. Cooper works at the famed Harvard Medical School and the equally renowned Massachusetts General Hospital. In 1978, he and his colleagues published their first article on medical error, "Preventable Anaesthesia Mishaps: A Study of Human Factors," in *Anaesthesiology,* the top journal in the field.[39]

To learn more about how medical error contributed to patient injury and death in anaesthesia, Cooper used the "critical incident technique" – the same method that Paul Fitts and Richard Jones had used to understand threats to aviation safety during World War II. Anaesthesiologists were asked to remember and describe incidents that either could have led or did lead to a bad outcome, which might be anything from adding to the length of the patient's stay in the hospital to permanent disability or death. The anaesthesiologists were then asked to recall the circumstances surrounding the critical incidents. These "incident reports" would then be used to provide other anaesthesiologists with a way to learn from experience, by understanding the reasons bad things had happened, or almost happened, and identifying problems with products (e.g., poor equipment design) or work systems (e.g., long work hours) – potentially lethal "invisible hands" that were threatening patient safety. This understanding provided a solid basis for making changes and thus reducing error. Because the critical incident technique had been used effectively in other industries, like aviation, Cooper thought it could be useful for health care too.

Seventy-two in-depth interviews were conducted over a two-year period with both more and less experienced anaesthesiologists, and a total of 359 critical incidents were identified and analyzed. One of the most interesting findings was the frequency of human error versus equipment failure. Only 14 per cent of the incidents involved failures of equipment, whereas human error was involved in 82 per cent of the critical incidents. As Dr. Jeff Cooper noted, "Although the incidence of outright functional failure of equipment was low, machines can have shortcomings or faults in design by their propensity to encourage human error. Indeed, a substantial fraction of reported incidents of human error seemed to be related to the design or organization of equipment or devices."[40] Cooper and his colleagues were fully aware of the implications of this finding. It provided a basis for "the application of human-factors principles to anaesthesia, following the example of successful applications in fields such as aviation."[41]

Taking just one example, Cooper found a problem with the design of the hoses and the various drug delivery ports on anaesthesiology machines. The hose for each drug was supposed to be connected to a different delivery port, yet any hose would fit in any port, making it possible and even likely that the anaesthesiologist would sooner or later unintentionally connect a hose to the wrong delivery port on the machine. This type of mistake – caused by a lack of differentiation in the hoses and delivery ports – would mean the patient might receive the wrong kind of drug, making lethal errors more likely.

Thanks to Cooper's studies, these and other design deficiencies were eventually eliminated using design principles that had been proven to work in other industries. Equipment manufacturers began to design the gas nozzles and connectors so that each connector could be plugged only into the correct nozzle; it wouldn't physically fit into any of the others – a behaviour-shaping constraint. These changes can be traced back to Jeff Cooper's seminal work, and the result was an enormous improvement in patient safety. A Human-tech Revolution in anaesthesia was underway.

Once we remove the blinders imposed by the Mechanistic and Humanistic approaches (or in the case of health care, super-Humanistic) a whole new world of connections opens up, and we can see that the amusing but effective flies etched onto the urinals at the Amsterdam airport do have something in common with an innovative life-and-death computer-based display in a nuclear power plant. From the binocular Human-tech perspective we can see that both designs make use of technical and human insights to achieve their purpose, and the end result in both cases is a tight fit between people and technology. Like yin and yang or a blind man and a cane, the two parts of the system work together as one indivisible whole – a harmonious system. And on top of that, we get technology that solves a worthwhile problem rather than rudderless technology for its own sake.

Many of the examples I've described seem almost like common sense. But the fact that the world is littered with so many complex technological systems that don't have any affinity with human nature shows that Human-tech thinking isn't nearly as prevalent as we might expect or want. In fact, it's quite rare. How can something that seems like common sense be so scarce? Again, Arthur Koestler provided some insight: "It all looks beautifully obvious – in the rear mirror. But there are situations where it needs great imaginative power, combined with disrespect for the traditional current of thought, to discover the obvious."[42] It's only after we've achieved a Human-tech Revolution and escaped from the clutches of the "traditional current of thought" – the one-eyed Mechanistic and Humanistic perspectives – that the power and elegance of designing technology to fit with human nature become "beautifully obvious."

All of this is reassuring and gives us hope that perhaps we can use technology to improve our quality of life, to move beyond the transitional instability that we're experiencing. But as technological systems become larger and more complex, the demands that they impose transcend the characteristics and the needs of the individual. The human body and the human mind are important; but there's more to human nature than just the physical and psychological attributes of a single person. What about the broader, "softer" elements of technological systems – teamwork, organizational structures and politics? As it turns out, the Human-tech approach at these levels, and the benefits to humankind, are even more vital.

6

Staying on the Same Page: Choreographing Team Coordination

THE HUMAN-TECH LADDER

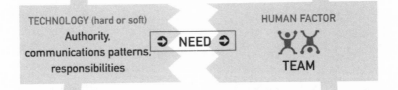

TECHNOLOGY (hard or soft)	HUMAN FACTOR
Authority, communications patterns, responsibilities	⊃ NEED ⊃
	TEAM

Anyone who has ever been involved in even a fleeting relationship knows that when you bring two people together you get a new emergent phenomenon – in fact, a new system – that is made up of more than the sum of its parts. Bring more than two people together into a team, or group, and the phenomenon becomes more complicated still. Of course, the behaviour of a team is still affected by the factors that shape the behaviour of each individual team member, but there's an additional and entirely different level of considerations that affects the behaviour of the team as a whole. When you're by yourself communication is a non-issue, unless you enjoy talking to yourself. Similarly, when you're working alone, the task of dividing up work responsibilities and figuring out who should be in charge is quite uncomplicated. All this changes when you have a team. Key factors such as communication, authority, responsibility and priority-setting must all be taken care of, otherwise the team members won't be able to coordinate their respective actions. A viable team relationship can't be based solely on independent, selfish behaviour. We learn this early in our family interactions, and it's essential for good managers to recognize this if office life is to run smoothly and the business or organization to succeed.

The Human-tech ladder tells us that these lessons are critical for effective system design. A viable complex technological system can't be developed merely by fitting the design to the physical and psychological characteristics of the individual team members. Affinity with human nature must be achieved at the team level as well. Designers must create a system that is tailored to the characteristics and needs of the team as a distinct entity in its own right. If they don't, the system won't run effectively and accidents will occur. The commercial aviation industry was the first to learn this lesson, being the forerunner in applying the relational approach at the team level, although here, too, the lesson came with a tremendously high price tag. Only after many costly accidents that led to destroyed airplanes and many lives lost did the importance of designing the technological systems to fit people working at a team level become clear.

The lessons themselves were often heartbreakingly mundane and human.

THE PESKY LIGHT BULB:
UNINTENTIONALLY FLYING INTO THE FLORIDA EVERGLADES

Late on the night of December 29, 1972, three experienced and able-bodied professional pilots in charge of Eastern Airlines

Flight 401 flew a Lockheed L-1011 with 176 people onboard *into* the Florida Everglades.[1] Ninety-nine people were killed. The airplane hadn't experienced any mechanical failure. The weather wasn't out of the ordinary. There was no act of sabotage or terrorism.

When Flight 401 was on its final approach to Miami International Airport, heading for runway 9 Left, the pilots flicked some controls in the cockpit to lower the landing gear, and something unexpected happened. A small indicator light should have come on after the nose landing gear had been lowered and locked into place. Contrary to the pilots' expectations, the light didn't come on this time. That meant one of two things – either an equipment failure (a mechanical problem with the landing gear) or an instrumentation failure (as simple a possibility as the light bulb being burnt out). The pilots suspected the latter was more likely, but they weren't sure. The light bulb could be manually pulled out of the cockpit panel to see if it was burnt, so being good problem-solvers, that's exactly what the crew attempted to do.

The only problem was that everybody in the cockpit got into the act at the same time. The captain, first officer and flight engineer all tried to figure out if there was a problem with the little light bulb, and nobody bothered to keep on monitoring the cockpit instruments. Flight 401 had been flying at 600 metres with the autopilot on, but at some point, the autopilot became disengaged, and the plane was no longer being flown by the automation.[2] None of the three crew members noticed; they thought that the automation was still in command of the airplane's flight path. With nothing controlling the plane – human pilot or autopilot – Eastern Flight 401 slowly started to descend. Given that the plane was already close to the ground, the situation was disastrous.

Thanks to the voice cockpit recorder and a television reenactment of the accident, we know what the crew members said to each other as their airplane was approaching the ground.

Flight Engineer: "Do you want me to test the lights or not?"

Captain: "Yeah. Check it."

First Officer: "It could be the light. Could you jiggle the light?"

Captain: "It's gotta come out a little and then snap in."

Captain: "See if you can pull that light out . . ."

First Officer: "You got a handkerchief or something so I can get a little better grip on this? Anything I can do it with. This damn thing just won't come out. If I had a pair of pliers I could . . ."

First Officer: "It's gotta be a faulty light. This damn thing just won't come out."

While the crew's collective attention was fixed on the light bulb, the plane kept descending – until it crashed into the earth in the Florida Everglades. Unbelievable as it seems, nobody was minding the store.

That's how Eastern Flight 401 wound up being another tragic headline. But there was one other small – but critical – detail that gave some insight into how this kind of disaster could be prevented. About eight seconds before the plane crashed, the recording makes clear that one of the crew members did notice that the autopilot had become disengaged, that the plane was descending, and that it was at a very low altitude. Eight long seconds. Enough time to save the day. All that crew member would have had to do was grab the flight controls and yank the plane up out of danger. Yet, the cockpit data recorder showed that the pilots didn't touch the controls during those final eight seconds; there was no last-ditch effort to climb. They were so out of touch with the primary task of flying the airplane that they were incapable of assimilating what was taking place.

Think about the Fender Strat design we looked at earlier. Because the design of the guitar was tailored to fit the player's body and other physical capabilities, the instrument and musician become one balanced and tightly integrated system – that harmonious yin and yang. With experience, the guitar becomes an extension of the human body; the guitarist doesn't consciously think about the properties of the guitar at all – he or she just focuses on the job to be performed, producing beautiful music. In a sense, the guitar becomes "invisible" to the player.

The interaction between the crew of Flight 401 and the cockpit technology was the exact opposite. Instead of the technology being invisible, so that the crew's attention could be focused on the task at hand, the task of flying the plane became invisible, because their attention was focused solely on deciphering what was going on with the perplexing technology. How did this happen?

A lack of effective teamwork was the key factor. Many commercial airline pilots started their careers in the military. The traditional attitude of the stereotypical jet fighter pilot with the "right stuff" may be exaggerated, but like many stereotypes it has some basis in fact. These pilots were taught, and perhaps even forced by circumstances, to be individualistic. They weren't trained to work together as a team. The "right stuff" attitude didn't allow them to admit to difficulties – asking for help would be unmanly. Such a culture made it unlikely that pilots would develop good team skills.

Aviation researchers have identified two frequently encountered patterns of leadership failure.[3] The crash of Eastern Flight 401 illustrates, in a tragically memorable way, what can happen when a captain doesn't take an authoritative leadership role by setting priorities, delegating tasks, and making sure that all the important responsibilities are covered; because the captain didn't take charge and organize the team, everybody tried to deal with the problem of the light bulb and nobody monitored the instruments – a deadly mistake. A second pattern of leadership failure occurs with a strong, autocratic leader who doesn't listen to others. This was what happened when a United Airlines DC-8-61 flight crashed near Portland, Oregon.[4] The accident investigation discovered that the macho attitude of the captain had intimidated the crew members –

which resulted in a lack of teamwork and eventually a fatal accident. The DC-8 was approaching Portland International Airport when the crew experienced a landing-gear malfunction. They circled near the airport for about an hour in an attempt to buy time while they coped with the problem. The two more junior crew members noticed the seriously low fuel level, but were too fearful of their authoritative captain to bring it to his attention, and the plane ran out of fuel. All four engines failed and the plane crashed about ten kilometres from the airport. Ten people were killed and twenty-three injured.

To be "in charge" is to take on a complex responsibility. We experience it daily – if we drive a car, for example – but the situation is even more demanding when other people's lives are in your hands. The kind of people who gravitate to such positions of authority generally have quite healthy egos, and the unassertive behaviour of first officers and flight engineers is understandable when their captain makes it clear that he (usually "he") dislikes being questioned or criticized – and in extreme cases simply won't tolerate it. The following anonymous incident report provides a blatant case in point:

> I was the first officer on an airline flight into Chicago O'Hare. The Captain was flying, we were on approach to 4R getting radar vectors and moving along at 250 knots. On our approach, Approach Control told us to slow to 180 knots. I acknowledged and waited for the captain to slow down. He did nothing, so I figured he didn't hear the clearance. So I repeated, "Approach said slow to 180," and his reply was something to the effect of, "I'll do what I want." I told him at least twice more and received the same kind of answer. Approach Control asked us why we had not slowed yet. I told them we were doing the best job we could and their reply was, "You almost hit another aircraft." They then asked us to turn east. I told them we would rather not because of the weather and we were

given present heading and to maintain 3000 ft. The captain descended to 3000 ft. and kept going to 2500 ft. even though I told him our altitude was 3000 ft. His comment was, "You just look out the damn window."[5]

Such an intimidating, overbearing attitude can make crew members decide to keep quiet rather than question the authority of their captain – even to the point of risking a deadly crash, as happened in the case of the United Airlines DC-8 near Portland.

These two tragic accidents are, sadly, not isolated examples. Industry reviews in the 1980s showed that over 70 per cent of aviation accidents resulted from poor crew coordination and communication.[6] Even with mechanically sound aircraft and crew members in good health and well trained in the technical "stick-and-rudder" flying skills, there can still be an "invisible hand" pulling the crew toward disaster. As one researcher aptly put it, individual pilots don't crash airplanes, flight crews do.[7]

This simple insight had tremendous implications for system design when the aviation industry took on the task of improving safety. The researchers and engineers realized that better engines and stronger wings weren't going to make flying much safer, because in most accidents there was absolutely nothing wrong with the engines, the wings or any of the other mechanical components. Fitting the system design to individual human physical and psychological characteristics wasn't going to go very far toward improving safety either, because in most accidents the problem wasn't in that area. The trouble was elsewhere. Planes were crashing and people were dying because of a *systems* flaw – because of the way in which crews failed to communicate and coordinate. Here was a significant societal need that had to be addressed.

TEACHING TEAMWORK: COCKPIT RESOURCE MANAGEMENT

It's worth taking a look at just how the aviation industry went about solving these problems, because the lessons learned there can and should be followed by other safety-critical sectors – such as the health-care, maritime, railway and nuclear sectors. From our Human-tech perspective, the objective is clear enough: to be effective, the technological system must pass the test of affinity with human nature *at the team level.* Exactly how this aim could be achieved by the aircraft designers in the 1980s was not so obvious. One option was to physically redesign the cockpit to facilitate more effective coordination. But it didn't seem likely that this approach would fix the problem because the crew members were already sitting right next to each other. Substantial changes to the physical layout of the cockpit were unlikely therefore to improve coordination further.

A very new and different idea was to change the way pilots were trained. At first blush, this might seem to contradict the idea of tailoring the design of technology to human nature, because the thing being designed isn't a physical object. After all, a training program is not a piece of hardware. But the definition of "technology" that I've adopted includes the "softer," non-physical kinds of system design, and crew training, which is one such "soft" aspect, has a direct bearing on aviation safety. Its impact illustrates why this broad definition not only makes sense but is essential, given the increasing complexity of modern systems.

First, although it doesn't result in the creation of a physical object, the development of a training program is a real and necessary part of system design, just as much as the design of equipment. In any complex technological system, the system design isn't complete unless and until the training program is specified.

In fact, in aviation, the content of training programs is strictly regulated by law. You can specify the technical characteristics of the wings, the engines, and other hardware components in as minute detail as you want, but the airplane simply won't be allowed to get off the ground unless the cockpit is staffed by licensed pilots who have passed the required training regimen. Training is part of the system's structure.

Second, even though training isn't a physical object, the content of the training program indirectly determines how well the crew interacts with the airplane – it creates relationships between the activities of the individual crew members, relationships that can lead either to tension or to harmony. The cockpit can be meticulously designed from a technical, physical and psychological perspective, but if the pilots aren't trained properly, they won't have the requisite flying or team skills for safe flight. In a complex technological system like aviation, you can't just throw people into the workplace and expect them to succeed. So it's not only legitimate but necessary to think about the creation of a training program as an equally important part of the overall design of the system, one that has a bearing on the ultimate impact of the technical components themselves.

So a Human-tech approach makes sense at the team level. Just as the size and shape of a toothbrush will allow people to brush their teeth effectively if it is consistent with what we know about human anatomy, the crew's communication patterns and responsibilities can be tailored to be consistent with what we know about team coordination, enabling crews to fly airplanes safely. In both cases, the overall aim is to satisfy our human or societal needs by tailoring the design of the system to fit human realities.

People first began putting these ideas into effect in the 1980s, when Earl Weiner, Robert Helmreich, Clay Foushee and others in the aviation industry started to advocate a new kind of pilot training program called Cockpit Resource Management (CRM).[8] CRM was a breakthrough in team training – it complemented traditional training programs by teaching effective teamwork techniques, not just individual "stick-and-rudder" skills. CRM is a Human-tech success story that improved aviation safety, so it's worth taking a look at how it achieved its goals.

To begin, Weiner and his colleagues looked at the societal need that had to be satisfied – in this case, the need for safety in the air in the new conditions imposed by modern air travel, and society's need to feel they could depend on the human crew. Older airplanes were designed to fly with a crew of three people (captain, first officer and flight engineer), and the primary role of the first officer was to serve as a backup to the captain in case of emergency. As new types of airplanes were introduced, the crew size was reduced to two (captain and first officer). As a result, first officers are no longer just there as a backup for manually flying the plane. Instead, they share in most tasks that must be performed in a modern cockpit. As a result, team coordination and communication becomes critical.

CRM training, which addressed this need by teaching pilots to work together effectively, is still practised to great effect. Some of the key skills are taught in a traditional classroom setting; pilots learn to recognize the classic patterns associated with poor teamwork:

- a failure to delegate tasks and responsibilities
- a failure to identify and communicate priorities, intention and plans
- a failure to maintain the monitoring of basic flight instruments

Cockpit voice recordings of real accidents where these failures proved fatal are used to create an unforgettable memory of how crucial crew coordination is to aviation safety. As one industry insider put it, there's nothing quite like hearing the voices of a crew plummeting to their death to get the attention of a group of professional pilots.

In the classroom, pilots don't just learn what to avoid, they also learn what to do: how to take an active role, behaviour that will contribute to safe flying. They're taught that communication among crew members is essential to ensure that everybody is on the same page. All the crew members should know what the current problem is, what the plan is for solving the problem, and what they each need to do to implement the plan. This coordination ensures that everybody is working on the same problem, and that everybody has the same understanding of what the priorities are – what to watch out for, and when to perform certain activities. Crew members are also trained to state goals, acknowledge orders, offer suggestions, predict outcomes to make expectations explicit, ask for help when they need it, and offer help when it looks like someone else needs it.

This classroom instruction plays an important role, but the heart of CRM training is in the full-mission flight simulator. This is where pilots put their classroom studies to the test, by flying very realistic simulated missions that have been especially designed to require effective crew coordination.[9] Typically, the missions will involve a sequence of challenges, such as bad weather, followed by equipment failures, or confusing changes in air-traffic-control instructions. The realism is what makes the simulator training so important.

I first saw an airplane simulator when I visited NASA Ames Research Center in Moffet Field, California. All the instruments and controls look like the real thing. They're in exactly the same places as they would be in a real airplane cockpit. They're labelled in the very same way. The pilots' seats are just like the real seats. And thanks to the magic of computer graphics the visual scene outside the cockpit windshield is brilliantly simulated: the night-time view is particularly compelling; to me, at least, it was indistinguishable from the real thing. Communication with air traffic control, cabin crew and ground crew is also simulated. But the best part of the simulator – the most fun – is its moving base.

The cockpit is on a platform connected to hydraulic lifts, so it moves. If you grab the control yoke and pull up, you feel as if you're being lifted up in the air, and if you push down sharply on the yoke, you have the sensation of plunging toward the ground. If the simulator is programmed to simulate flying through turbulence, you're in for a bumpy ride. A modern cockpit simulator is like one very big, very exciting high-tech video game costing millions of dollars.

But the goal isn't entertainment; it's to teach crews how to save lives when problems occur in the sky. The entire session is videotaped, and after the simulated flight, a debriefing session is held, during which the crew watches and critiques their own performance with the help of an expert facilitator. They can see exactly what they did, what everyone else did, what they said, and what happened during each stage of the simulated flight, identifying both good and bad patterns of coordination. But the focus is always on critiquing the behaviour, *not* the person. After the review session the tapes are erased, so the crews can be sure their training performance won't ever be used against them.

By all accounts, the experience of reviewing a challenging session in a realistic simulator is invaluable in hammering home the CRM lessons the pilots covered during their classroom training. Some pilots are astounded when they see themselves. A captain might realize, for example, how domineering he is: "Am I really that aggressive when I give out orders? Holy cow, I had no idea. No wonder my first officer doesn't speak up more often." A crew can see exactly when and how their communication failed, and how their coordination suffered as a result: "At that point in the flight, I was really busy but I didn't ask for help. Jesus, I was trying to do three things at the same time. I really should have said something. A helping hand would have made a big difference." Because the scenarios are deliberately designed to teach CRM principles, the crews don't just learn the importance of effective teamwork in the abstract, as they would in a classroom. They learn it in a tangible, personal and therefore memorable way. If you've ever watched yourself give a speech on videotape, for instance, you'll know exactly what I'm talking about. It's an experience that's hard to forget.

You might think that more sophisticated computer technology might reduce or even do away with the need for CRM, but the problem of crew coordination is actually made *more* complex in the latest cockpits, with their new array of controls for computer automation (and not just because there are now only two crew members instead of three).[10] Old-style cockpits had analog meters, knobs, switches and other types of controls spread out all over the cockpit panels. As a result, the physical movements of one crew member were a visible indication of what he was doing: if you saw your fellow crew member leaning to the right and putting his hand on a certain switch, you knew he was probably raising the landing gear, but if he was looking upward at an analog meter, you knew he was probably checking on the status of the electrical systems. In the old-style cockpits, one crew member could frequently even guess what the other member was thinking about. Information was freely available. All you had to do was glance out of the corner of your eye and you could observe what your partner was doing. And most important of all, this "free" information was extremely useful for crew coordination. If a quick look could tell you what your fellow crew member was working on, it was easy to synchronize your actions. Like a pair of ballet dancers, a crew could coordinate their actions silently, with no words spoken or questions asked, merely by glancing at each other's movements. The crew-and-airplane could become one.

The arrival of computer technology in the modern cockpit changed all that – inadvertently. Now, almost all the information is presented on computer monitors, and thanks to the marvels of automation, pilots can bring up displays for checking on the hydraulics, displays for looking at the weather, and so on; it's all at their fingertips – they just have to choose which display to look at. As a result, pilots don't have to reach for knobs and dials or move their heads to look at different instruments nearly as much as they used to in the older-style cockpits.

And therein lies the problem. Because physical movement is curtailed, pilots don't receive as much information from the actions of their partners. Most of the time, they're just sitting and staring at the computer screens, regardless of what they're thinking about or working on. So there isn't as much "free information" as there used to be; each crew member has to explicitly communicate his or her intent and actions by talking to the other, if they're to keep in tune with each other. That takes a lot more effort and concentration than a glance out of the corner of your eye. Instead of a pair of expert ballet dancers, crews in a modern cockpit risk looking more like a pair of clumsy (and talkative) teenagers struggling to tango. In the extreme, the two-person crew and the airplane can become three separate entities. The potential for bad fit is huge.

So CRM training is actually more, not less, important in the newest highly automated cockpits, because the reduction in free information must be compensated for by even more explicit efforts to delegate responsibilities, set priorities, update other crew members, and so on. Technical advances don't do away with people or the need to incorporate human nature into the design of the system. Concepts like CRM training are more relevant and more critical than ever before.

CRM is a Human-tech success story, showing that it is both necessary and possible to design for human nature at the team level in complex technological systems. Commercial airline statistics show that CRM training improves safety, and a substantial reduction in accident rates after CRM training has also been observed in military aviation.[11] Several pilots who have faced emergency situations cited their training as having helped them cope and save lives. The benefits are so widely recognized that the failure of an airline to offer CRM training has even been used in court as evidence of negligence on the part of the airline.[12] It is now legally mandated for all U.S. military services and commercial carriers.

These achievements are particularly impressive if we consider that aviation is a conservative industry that doesn't accept change lightly. Training programs are expensive, so advocates for change had to provide convincing arguments to show that the benefits of CRM training were worthwhile, especially since adoption occurred during a time when the industry was in bad shape economically.[13] But the potential benefits to the aviation industry were simply too compelling to be denied.

WRESTLING IN THE O.R.: TEAMWORK IN MEDICINE

Worcester, Mass. Nov. 27 (AP) – A state medical board has fined a surgeon and an anaesthesiologist $10,000 each for brawling in an operating room while their patient slept under general anaesthesia. . . . In addition to imposing the fines, the state board of Registration in Medicine last week ordered the doctors to undergo joint psychotherapy. It also directed officials at the Medical Center of Central Massachusetts, who had already put the doctors on five years' probation, to monitor Drs. Chan and Korgaonkar for five years. The medical board said that on Oct. 24, 1991, Dr. Korgaonkar was about to begin surgery when he and Dr. Chan began to argue. . . . Dr. Chan swore at Dr. Korgaonkar, who threw a cotton-tipped prep stick at Dr. Chan. The two then raised their fists and scuffled briefly, at one point wrestling on the floor. A nurse monitored the anaesthetized patient as the doctors fought.[14]

Although aviation was the first safety-critical sector to adopt team simulation training to foster coordination, pilots are not the only ones who work in teams or groups. Health-care professionals are also required to work together in the treatment of patients – in an operating room, for instance. As the astounding incident involving Drs. Chan and Korgaonkar suggests, the coordination problem is arguably even more daunting in health care than it is in aviation. The men and women who make up a crew have a similar background, the primary difference being their levels of experience and seniority. Both the pilot and the first officer are capable of flying the plane. But teams in health care are usually far more diverse: many surgeries require a team comprising a surgeon, an anaesthesiologist and nurses. A surgeon and an anaesthesiologist are both doctors of medicine but their specialist training after the initial degree makes their areas of knowledge quite distinct. Neither is qualified to do the other's job, making it all the more essential that they know how to work together.

That's where the potential for conflict begins: surgeons have a legitimate basis for seeing themselves as team leaders – but so has the anaesthesiologist.[15] The purpose of surgery is to operate on the patient, and it might seem as if that criterion alone would justify the surgeon in thinking he or she should be in charge in the operating room. But managing the patient's life signs is not a secondary consideration, and that being the anaesthesiologist's domain of expertise, anaesthesiologists have an equally solid basis for believing they should be in the driver's seat. These tensions, if they get out of control, can seriously impair the effectiveness of the team.

The team-coordination problem is exacerbated in the O.R. because surgeons and anaesthesiologists have competing criteria for starting their work. Anaesthesiologists want to delay administering anaesthesia for as long as possible – ideally, until just before surgery begins – so they don't have to monitor the unconscious patient for long periods while waiting for the surgeon to show up. In contrast, surgeons want to have anaesthesia administered as soon as possible so their time will not be wasted; they want to start working on the patient right after they walk into the operating room. Since scheduling surgeries is far from an exact science, these conflicting criteria can cause tempers to flare. I never witnessed an actual brawl in an operating room during my research in health care, but I have seen surgeons react with irritation when a patient wasn't sufficiently anaesthetized to allow the operation to begin, as if to say: "Get your act together! How dare you waste my valuable time?" I've also seen anaesthesiologists in operating rooms look at surgeons with equal contempt after they walk in late without offering an explanation or apology, as if to say: "How dare you keep us sitting around waiting, then walk in late without apologizing, and expect everyone to be at your beck and call? Being a surgeon doesn't give you the right to disrespect the rest of us."

But coordination in an operating room doesn't stop with problems of leadership and potentially conflicting objectives; it can also be hampered by perceived differences in professional status among the various medical specialists. There's a very clear hierarchy that all medical students learn about early on in their studies. Some surgical specialties enjoy a great deal of respect and prestige, and others are less highly regarded. Orthopaedic surgeons tend to be at the bottom because their skills purportedly rely primarily on the brute force required to manipulate bones, joints and tendons; they are the manual labourers of surgical society. Cardiac and brain surgeons tend to be at the top of the hierarchy because of their mastery of the precise skills required to perform intricate operations effectively on a consistent basis. This "class system" affects how people deal with each other; the level of respect and deference offered to an individual is determined by his or her place in the hierarchy. The problems this can create in an operating room are easy to imagine: those lower on the totem pole may be unwilling to question the judgement of brain surgeons, and brain surgeons may be less willing to listen to input from others.

These attitudes, which are strikingly similar to the military "top gun" fighter-pilot mentality, may seem misguided and counterproductive, to an unbiased observer. After all, every competent surgeon has a valuable set of skills; I wouldn't want an orthopaedic surgeon to perform brain surgery on me, but I also certainly wouldn't want a brain surgeon to reconstruct my ailing knee. Every skilled specialist deserves respect. But the medical profession, rightly or wrongly, has placed great emphasis on prestige, and whether these perceived status differences are justified or not doesn't really matter in the end because they're real and they're important to those involved. A recent field study conducted in operating rooms found that every single operation that was observed had between one and four "high-tension" events that strained communication between the team members.[16] And if you think fisticuffs and wrestling in the operating room are as bad as it gets, guess again:

> Rio de Janeiro. Reuters. A Brazilian surgeon shot a colleague, who was responsible for the anaesthesia of the patient, during abdominal surgery. While this was happening . . . the patient awoke from anaesthesia and, on seeing the bloodbath, fainted. The Resident who was present attempted to save the life of the anaesthetist, then ended the abdominal operation. The surgeon was long gone over the mountains. There was disagreement regarding the surgery between the two doctors, members of a private clinic at Macae, near Rio de Janeiro, where the operation took place. During the dispute, the 60-year-old surgeon, Marcelino Pereira da Silva, took out a revolver and put three shots into the head of Elimson Ribeiro Elias, age 40. Search is on for the surgeon.[17]

John Doyle, an anaesthesiologist colleague of mine, once told me that good anaesthesiologists have to check their egos at the door, since status conflict with the surgeon can threaten coordination in an operating room, and potentially, the patient's safety. He knew of an anaesthesiologist who, as his first action in an operating room, would deliberately turn the patient monitoring equipment just enough to place it out of view of the surgeon, thereby tilting the balance of power in his favour by creating a situation where the "hotshot" surgeon was forced to be dependent on him to obtain information that was critical to performing the operation successfully. John's view is that such power struggles have no place in the operating room. If necessary, he would rather defuse potential flare-ups by letting the surgeons feed their egos, letting them be in charge, thereby minimizing conflict and improving patient safety. In the absence of a better solution, that is an effective way of handling such a moment, but it's far from ideal.

Such situations may seem alarming, but I haven't even begun to describe the challenges faced by one of the most frequently ignored yet critical professions in health care: nurses. If some brain surgeons look down upon orthopaedic surgeons, just imagine how they view lowly nurses, who don't even have an MD. Nurses are too often perceived to be off the scale at the bottom of the health-care hierarchy. Yet they play a unique, invaluable role. For starters, they're the ones who generally have the most contact with patients. They get to know their names without having to consult a patient chart, they become familiar with the patient's family; outside the operating room, they're often the ones who deliver health care; physicians give orders, but nurses implement them. Nurses provide the vital human touch that is frequently missing in physicians who've been trained to be dispassionate health-care providers. Yet many physicians wouldn't dream of taking advice from a nurse, and not surprisingly, many nurses are hesitant to offer their point of view because they know the kind of reaction they're likely to get.

Admittedly, wrestling and gunfights in the operating room are extremely rare. I've deliberately painted a pessimistic – some would argue, exaggerated – picture of the tensions that can threaten team coordination in health care. There are many anaesthesiologists and surgeons who get along well, there are many physicians who not only listen to nurses but actively seek their input, and there are many operating room teams that work fluently and efficiently, like a well-oiled machine. Nevertheless, the tensions I've described between health-care providers are real, and because they can affect patient safety, it's important that they be openly recognized. Just as aviation has found ways to deal with the "top gun" attitude, health care should deal with the threats to coordination posed by the conflicts I've described.

Health care has lagged behind the aviation industry in applying Human-tech thinking in situations that require teamwork; and while this delay undoubtedly has had a negative impact on patient safety, it did have one beneficial side effect. By the time the health-care community realized the importance of coordination in health-care system design, the successful role model of CRM was available for study and for adaptation to health care – which is exactly what happened. And once again, it was the anaesthesiology community that led the way in initiating Human-tech thinking into health care.

THE SMART DUMMY:
PATIENT SIMULATORS FOR TEAM TRAINING

The seminal work applying CRM techniques to the operating room was published by three anaesthesiologists – David Gaba, Steve Howard and Kevin Fish – and their colleagues at Stanford

University in 1992.[18] The rationale behind their work sounds simple: if we can train pilots in an airplane simulator to develop their team skills, why not train medical professionals in an operating room simulator to enhance their team coordination skills? This was an idea that had to be worked out in several stages. It wasn't possible to simply transfer the approach that worked so well in aviation to the health care environment. It had to be adapted for its new purpose. CRM-like training in health care couldn't get off the ground without the medical equivalent of a modern flight simulator.

So the first order of business for the Stanford team was to invent a realistic simulated "patient." Their design has since gone through several revisions, and is now available as a commercially available product.[19] I've seen several of them in operation, and I'm continually impressed by the level of realism they can re-create. The centrepiece is a "dummy" that looks like the mannequins that you see in first aid classes. This "dummy" is smart, however. It's fully instrumented and connected to two computers that mimic how the human body responds to different kinds of events. The mannequin has a built-in drug-recognition system that can simulate how people react to different types of medication, so during the training scenarios, participants can administer real drugs to the patient in the same way that they would in a real operating room, and the patient's vital signs change accordingly on the monitoring equipment, in real time.

This feature alone is an extraordinary technical achievement, but the patient simulator has many other equally impressive features. For instance, the mannequin's eyes are computer controlled so they can open and close in response to external stimuli. The mannequin's electromechanical lungs are also computer-controlled and thus breathe either spontaneously or in response to mechanical or hand ventilation; the arms can move in response to stimuli as long as the patient isn't paralyzed; the arms and legs can become swollen to simulate trauma. The simulator also re-creates the breathing and heartbeat sounds of a human patient. There are even electronic leads where a real ECG machine can be connected to assess the patient's state.

Behind the scenes, out of view of the training participants, the computers controlling the mannequin can simulate more than twenty abnormal events, of varying degrees of severity, traceable to different triggering conditions, and escalating either rapidly or slowly. Up to three abnormal events can be running at the same time, creating an almost endless set of unique scenarios from which to choose. And just to make sure that the nurses and doctors deal with these events in a realistic way, the mannequin lies on an operating room table and is connected to real clinical equipment to monitor its life signs. The whole setup is as close to "real life" as it can be.

When the simulator is in use, there are several participants, each playing a different assigned role. One person acts as the anaesthesiologist, another as the surgeon, another as a nurse, and so on. People can be rotated through the various roles so that all members of the team will understand what the others are doing: surgeons find out what it feels like to be a nurse, and an anaesthesiologist can see what the surgeon sees, for instance. All the participants are given information about the patient and the kind of operation that's being simulated (e.g., sixty-year-old male undergoing double bypass surgery). Everyone does what they would do in an actual operating room. There's even a speaker that's connected to a microphone in a "control room" outside the simulator, where the person who's running the simulator sits. That person can talk as if he or she were the patient, can watch and hear what everyone in the simulator is doing and saying, and enter commands into the computer simulator to create difficult situations for the participants to handle. For example, the simulator controller may want to send a command to the computer to simulate a heart attack to see how well the operating room team can work together in the face of an unexpected emergency.

The creation of the patient simulator was obviously their biggest challenge, but with that "minor detail" out of the way, the Stanford team faced another complex task: developing an effective training curriculum. The teaching objectives were to be similar to those in aviation: members of a team learn to exercise effective leadership, ask for help early on, communicate effectively, distribute the workload evenly, use all available human and equipment resources, and so on. However, as I mentioned earlier, some of the threats to effective coordination in operating rooms are markedly different from those in cockpits; the training model couldn't just be transferred from aviation to medicine without modification. The Stanford team had to develop a brand new curriculum tailored to the unique challenges of team coordination in health care. This stage of the process was dealt with as effectively as the first. The Stanford team developed a textbook that is now required reading for students in CRM-like training courses in anaesthesiology.[20]

There are now various CRM-like courses in health care offered in patient simulator centres all over the world, including two in Toronto, where I work. I was lucky enough to be able to experience both the training program and the simulator, first hand, by being a participant in a session at one of the Toronto sites. In one of the test scenarios, I played the role of a scrub nurse who was supposed to hand instruments to the surgeon as he requested them. Since I wasn't deeply involved in the action, I had an opportunity to watch the other participants closely.

Things started going badly after the simulated patient experienced a complication. All of a sudden, the anaesthesiologist became extremely busy. He had to call upon all his expertise and decision-making skills. But the demands were too much for him to handle as the patient's state became critical. He called in another anaesthesiologist to help him save the patient's life.

The entire event was videotaped, and afterwards we all went into a debriefing room to watch the tape and see what we could learn. As we sat down, I glanced at the anaesthesiologist. He was visibly shaken. He was collapsed in his chair, taking deep breaths. He looked winded, as though he had just come back from running a ten-kilometre race. At first I was puzzled; maybe he was just out of shape? But as we watched the videotape and the anaesthesiologist explained what had been going on in his mind during the various stages of the scenario, I began to understand that his distress had been caused not by physical fatigue but by his emotional response to the realism of the simulator and the drama of the situation.

When he first went into the operating-room simulator, the anaesthesiologist said, he knew it was all fake and contrived – he was even sceptical that a setting as complex as an operating room could ever be simulated adequately. But as soon as they began to work and the patient's life signs on the monitor started to change in a realistic way, he forgot that this was not a real operating room with a real patient and he was caught up in the unfolding events. And he frankly described to us how, when the patient's vital signs suddenly deteriorated, indicating a life-threatening crisis, he had experienced the same distraught feelings as those he had felt before in a real operating room when he feared he was truly losing the patient. I'll never forget the look on his face as he reconstructed for us the sequence of thoughts and emotions that went through his mind as he desperately searched for ways to save the patient's life. The fact that the patient was a dummy and the operating room was simulated did not cross his mind after the first few minutes of the test scenario.

In another simulation event, I learned more about how that staged realism could be put to effective use in teaching coordination skills. This time I got to play a more active role in the story; I was told to behave like an overeager medical student and bombard the surgeon with irritating questions throughout the surgery. Partway through, an emergency was simulated by lowering the patient's vital signs. Ideally, the surgeon should have taken charge of the situation and delegated responsibilities to each of the team members. My job was to distract him. If he had learned how to function effectively as a team leader, he would have told me to be quiet when the emergency started so that he could focus on the important task at hand. Instead, the surgeon reacted to events rather than showing leadership. While the simulated patient was getting closer to dying, I pestered him with trivial questions like, "Hey, what does this thing over here do?" I felt guilty because the poor guy was completely frazzled, trying to answer my questions while a (simulated) life-threatening situation worsened. He should have told me to get the hell out of the way, but he didn't handle the situation very well at all – the simulated patient wound up dying. But that's the beauty of simulation. Better to learn the valuable team coordination lessons on simulated patients than lose real ones.

Because we're all consumers of health care at some point, we can all benefit when designers lend a helping hand, using every beneficial intervention they know of to help health-care providers do their very difficult job as well as they possibly can. Thanks to the creative and groundbreaking work of the Stanford group and others who have since followed their lead, we know how to choreograph fluid team coordination by building in harmonious relationships; Human-tech thinking – designing systems to fit people at the team level – is now a reality, although still primarily in anaesthesiology. In fact, simulator-based training has been so successful that it's now part of the anaesthesiology curriculum at several major universities, and programs are also offered as continuing education for medical professionals who weren't offered teamwork training in medical school. At Harvard, doctors who have gone through simulator training even receive a 10 per cent discount on their insurance premiums. Other medical specialties have started to follow the anaesthesiologists' lead, developing team training programs for intensive care units, emergency departments and trauma centres, delivery rooms, cardiac arrest response teams, and radiology units.[21]

But – and this is a big "but" – CRM-like training is still not *legally* mandated in health care, as it is in aviation, despite its obvious benefits to public health. A great deal of cultural change has to happen before Human-tech thinking really takes off in health care. But there are reasons to be optimistic. I'm confident that such training will eventually become a routine part of education in all sectors of health care, because there is so much to be gained. Helping people "stay on the same page," by design, makes so much sense that its value to society will surely come to be appreciated by all specialties in the health-care sector.

A GLOBAL SPIN:
CULTURAL EFFECTS IN TEAM COORDINATION

Because of the pioneering work in aviation, the Human-tech approach to system design at a team level is starting to be noticed in various other safety-critical sectors. For example, researchers have begun creating simulator training programs to help nuclear power plant crews and maritime crews communicate and coordinate more effectively.[22] These efforts are likely to become more common in the future as more and more industries respond to pressures in the changing workplaces of the twenty-first century.

But there's an interesting new wrinkle: anyone who has lived in more than one country knows that differences between nations can be dramatic, but what most people don't realize is that these economic, cultural and social differences may affect how successfully people work together. In aviation, researchers have found that cultural factors can have a direct correlation with safety: crew-related accident rates can be up to eight times more likely in one country than in another.[23] There are undoubtedly many reasons for these differences, but one interesting possibility is that cultural differences may affect whether crews are able to adapt to the principles underlying CRM training.[24]

In some cultures, status counts for so much that people find it hard not to bow and defer to authority, even in cases where the authority is clearly wrong or would benefit from another opinion. In the cockpit, this kind of cultural behaviour can lead to situations where subordinates are unwilling to question the authority of the captain, even if deadly disasters may result.

The opposite problem can be found in cultures that put a high value on individualism. In these countries, people sometimes put too much emphasis on their own assessment of a situation, discounting or even ignoring the input of others. A concern for personal prestige, image or other ego factors can override the unity of a group. The implications for aviation safety in those cultures that place a premium on individualism are significant because egotistical tendencies can make captains unwilling to listen to subordinates, taking even justifiable questions or concerns as a challenge to their authority. A desire to save face can override safety precautions, which can lead to fatal crashes by restricting the flow of communication and so lowering the crew's performance.

The findings show that cultural differences can play an important role in CRM training. And if cultural effects can have an impact even when all the crew members are of the same nationality, just imagine how much the possibility for conflict and misunderstanding increases when crew members come from different countries, and therefore, different cultural backgrounds. As globalization increasingly makes the world a smaller place and requires people from different nationalities to work together, proactively designing complex technological systems that place human relationships at the centre will become even more important than it is now. We can do it – we already know how to create Human-tech systems that foster team coordination, and we know that doing so can help avoid catastrophic accidents and save lives. Our future lies in our own hands.

The path to globalization has seen our systems spiral into increasing complexity. Thanks to technological innovations in the telecommunications sector, people are now able to communicate over longer and longer distances in a shorter and shorter amount of time. As a result, organizations – both privately and publicly owned – are now far more complex and farther reaching than ever. In the most extreme cases – multinational governmental coalitions, large non-governmental organizations, or transnational corporations – the reach of the "invisible hand" can literally span the entire globe, making it more and more difficult for an organization to control its operations effectively. Moreover, because the demands people face on the job are changing faster, there's a greater need to cope with novelty and change by designing learning organizations, the next rung on the Human-tech ladder. How well we tailor the design of the "softer" part of technological systems to these human organizational challenges will have a tremendous impact on our ability to maintain control over our technical creations.

7

Management Matters:
Building Learning Organizations

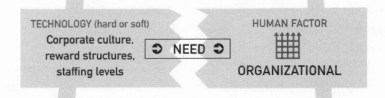

THE HUMAN-TECH LADDER

TECHNOLOGY (hard or soft)
Corporate culture,
reward structures,
staffing levels

⊃ NEED ⊃

HUMAN FACTOR

ORGANIZATIONAL

A few years ago, my mother went to see a surgeon about her varicose veins. She had already undergone three operations for this ailment over a thirty-year period. The surgeon examined both her legs and told her that, while both had problems, he recommended operating only on the left leg this time, because it was in worse shape than the right. A few weeks later, she went to the hospital to have the operation done. The surgeon started preparing her leg. Much to her surprise, however, he was working on the right leg, although he had said that he'd only operate on the left. My mother thought, Well, maybe he decided to do both and didn't bother telling me. I'm not going to say anything.

What do I know? I have to trust him. After all, he's a doctor. He went to school for all those years. He must know what he's doing.

But as the surgeon continued to draw on the right leg, she reconsidered. She decided she'd better say something, just to be sure. So she asked, "Are we going to do both?" The surgeon replied, "No, we're just doing one." My mother answered with increasing urgency, "We're supposed to do the left one, not the right. That's what you said in your office. You said you weren't going to operate on the right one." The surgeon, surprised, consulted the information on my mother's chart, and then said, "Oh, you're right!" He stopped preparing the right leg and moved to the left, as initially planned.

I never met my mother's surgeon, and I don't know exactly why this near miss occurred. Perhaps he was overworked from those crazy work hours that are a result of the "superhuman" demands so often made on health-care professionals – demands that ignore human limitations and expect people to be perfect, regardless of how imperfect their work situation may be. Or perhaps he was just distracted – the kind of momentary lapse of attention we've all experienced, albeit in more forgiving environments, but which happened to come at a particularly inopportune – and for my mother, critical – time. Because it's a well-known fact that all people, even surgeons, make mistakes and can sometimes be distracted, a number of hospitals have put in place a procedure with checks and balances – a helpful, safety-critical "invisible hand" – to help prevent such so-called "wrong-side surgeries."[1] Perhaps this surgeon's hospital didn't have such a procedure, or it failed. It's even possible that the surgeon was incompetent, although the scientific research shows that this is quite unlikely; more often than not, medical near misses are caused by qualified and conscientious people working in poorly designed systems.[2]

But regardless of the true cause – and there may have been several – shouldn't there be an organizational mechanism in place to take note of and follow up on near-miss experiences? It's only by identifying what went wrong – and why – and learning from errors that it will be possible to make systemic changes to prevent such close calls – or worse, wrong-side operations that actually carve into the relatively healthy leg of someone else's mother.

Designing learning organizations to keep track of what might go wrong can be used to identify and address many other threats to patient safety – for example, drug labels that look too similar to others and are easy to confuse, indecipherable prescriptions that are scribbled by doctors in a hurry, and so on. In one sense, the organizational level of our Human-tech ladder should be not so much an invisible hand as an invisible but vigilant overseer of all the lower levels, constantly on the lookout to remove bad fits between people and technology at the physical, psychological and team levels.

And the same idea can be applied outside health care. After the *Columbia* space shuttle accident on February 1, 2003, in which seven lives were lost, many people have been looking closely at NASA management to see if information that could have prevented the accident was available but not heeded.[3] History tells us that there are very good reasons for looking into that possibility.

A LATE-NIGHT CONFERENCE CALL: THE *CHALLENGER* ACCIDENT

At 11:25 A.M. on January 29, 1986, millions of TV viewers around the world watched as the final countdown for the launch of the Space Shuttle *Challenger* began.[4] On board was a crew of seven, including Christa McAuliffe, a teacher from New Hampshire who was going up in the shuttle to teach elementary school students from space. In addition to those watching from home, many people, including the families of the crew members, gathered that chilly morning at Kennedy Space Center in Cape Canaveral, Florida, to witness the historic moment.

Most of those watching had no idea that anyone was worried about letting the *Challenger* take off in such cold weather. The night before the launch, thirty-four managers and engineers from NASA and Morton Thiokol, the firm that had designed part of the shuttle, had participated in a three-hour conference call to discuss whether it would be safe to go ahead with the launch. There was a great deal of disagreement and heated debate about whether the O-rings – large, circular, rubber parts – on the shuttle would hold their seal at the unprecedented low temperatures that were expected the following morning. Some Thiokol engineers thought that the *Challenger* should not be allowed to take off the next day, but some NASA managers disagreed. For a while, it seemed as if the dissenters would prevail. In the end, however, the Thiokol engineers completely reversed their position and recommended, along with the others, that the launch go ahead as planned.

The next morning, *Challenger* left the launch pad, rocketing into the Florida sky. The temperature was 36°F (2°C). At first, everything seemed to go as planned, at least to the naked eye. Later, close-up images would show that small puffs of smoke had already started to escape from the O-rings while *Challenger* was still on the launch pad, an ominous but unnoticed indication that trouble was brewing. Seventy-three seconds after it began, the mission ended in deadly disaster as a fireball erupted fifteen kilometres above the Atlantic Ocean. Witnesses gasped with horror as *Challenger* became enveloped and completely obscured by an enormous cloud of smoke, and two of its parts broke off and spun wildly out of control away from the main explosion, creating a devastating TV image of tangled trails of smoke that would remain etched in our minds forever. None of the seven crew members survived.

After endless media scrutiny and a presidential commission report, the general public learned that organizational factors had played an important role in this technological disaster. Detailed analyses of the late-night conference call showed a subtle, yet critical set of interactions.

First, the burden of proof had shifted, without anyone realizing it: in the past, engineers had always been required to prove to their managers that it was safe to go ahead with the launch, but that night, the engineers were put in the position where they had to prove that it *wasn't* safe to launch – a complete reversal of procedure. Some critics attributed this about-face to financial, media and political pressures that caused NASA to get shuttle missions off the ground as soon as possible, perhaps unwittingly sacrificing safety for efficiency. Indeed, in response to the Thiokol engineers' initial recommendation not to launch, one NASA manager said, "I'm appalled," and another said, "When do you want us to launch, Thiokol, in April?" At one point in the conversation, a Thiokol senior manager was told: "It's time to take off your engineering hat and put on your management hat."

Second, there had been some early warning signs. Several shuttle missions had revealed "blow-bys" on the O-rings. These were black soot marks that revealed a partial breach of the seal – an unsafe condition. Not only that, but "blow-bys" were known to be more likely when the launch temperatures were colder (although there were exceptions, a factor that added "noise" to the pattern). The conference-call participants had access to some of this information, but given the shift in the burden of proof from the yes voters to the no voters, their review of the evidence wasn't enough to convince everyone that the launch should be delayed. An important opportunity to learn from experience – and thereby avoid disaster – was lost.

My main point here is not to second-guess the managers' decision to launch, but rather to show that organizational factors strongly affect the operation of complex socio-technical systems, and that considerations of expediency can lead to dangerous situations. The general lesson was captured in the immortal words of Richard P. Feynman, a Cal Tech physics professor and Nobel laureate who was a member of the presidential commission: "For a successful technology, reality must take precedence over public relations, for Nature cannot be fooled."[5] O-rings will behave a certain way at cold temperatures. This is a fact. Burdened managers may talk themselves into believing the risk is small enough to be ignored, but the facts can't be escaped.

But the physical world isn't the only design factor that can be unforgiving; so can human nature. Quite simply, if we violate what we know about human nature when designing complex technological systems, we wind up paying a tremendous social cost. Therefore, organizational decisions shouldn't be based only on what we know about the physical world. They should also take into account what we know about human nature.

The *organizational* level of the Human-tech ladder introduces several new factors that can't be found at the physical, psychological or team levels. Here we find systems of incentives, disincentives, staffing levels, management structures, information flows across teams and organizational cultures. These are all features of organizations, not of individuals or teams. Decisions made at the organizational level can have a big impact on human behaviour at lower levels. For instance, management decisions regarding staff levels in a hospital can affect the mental workload experienced by individual health-care providers. And this top-down influence is precisely why the organizational level is so important to the functioning of a technological system. A design can be well tailored to the characteristics of the body, the mind and the team, but all that forethought can go down the drain if organizational considerations aren't also dealt with effectively. A classic example is the failure to allow for information to be shared between teams within an organization, a pathology that is well known to anyone who works in a large bureaucracy (or reads the *Dilbert* comic strip regularly). Each team can be doing an excellent job of coordinating its internal activities, and each individual within the team can be doing an outstanding job of performing the mental and physical tasks for which he or she is responsible, yet the organization as a whole can flounder miserably if the various teams pursue conflicting objectives. This potential for problems reinforces the generalized lesson of the *Challenger* explosion: a technological system won't succeed unless sufficient attention is paid to organizational issues, such as how decisions about safety are made in the face of outside pressures.

Interestingly, however, the *management* of technology has often been viewed as an activity separate from the *design* of that very same technology. Business schools educate their students to be competent managers, but the technical training that would enable them to understand the operation of a complex technological system isn't considered necessary. Engineering knowledge is seen as "nitty-gritty technical details," which overworked managers shouldn't have to concern themselves with – although it would help them to assess how, for instance, cutbacks might affect safety. Conversely, engineering schools educate their students to be competent designers, but their graduates often know next to nothing about the skills required to run the business side of a complex technological system – knowledge that they could use if they were called on to oversee the administrative logistics of a large engineering project or supervise a team of engineers effectively and efficiently. In a parallel compartmentalization, engineers view those matters as "just management issues" which shouldn't distract the technically inclined Wizards from their primary responsibility as designers. Management and engineering design are treated as if they were two entirely separate things.

But even a brief acquaintance with real organizations reveals that this dichotomous viewpoint is seriously flawed. When most engineering students first get hired, say by a software engineering firm, their jobs typically require strictly technical knowledge. However, within a few years, many engineers get promoted into management positions that require them to oversee the design or operation of technological systems. To perform these jobs well, engineers need more than technical skills; they also need effective management skills. But having had no real-life management experience and no management training during their university years, engineers frequently find themselves unprepared for the challenges they face. Jeff Skoll, co-founder of eBay and alumnus of the University of Toronto's Department of Electrical and Computer Engineering, donated millions of dollars to his alma mater to initiate an innovative dual-degree engineering/ management program for precisely this reason.

Rather than provide engineers with management training, some organizations have hired graduates from business schools to oversee the design or operation of technological systems. Because these individuals don't usually understand the underlying technology that they've been put in charge of and don't usually consult with technical experts, they have little choice but to apply standard management procedures, regardless of the industry they're managing. Organizations that use this context-free approach to managing technological systems – public drinking-water distribution systems, for instance – have an abysmal long-term performance record, revealing how important it is for effective management to have access to industry-specific technical knowledge.[6]

Clearly, the design of technology and the management of technology aren't two separate things after all. In fact, I believe that management is an indispensable aspect of technological system design itself. It's just as real and just as important as the creation of hardware and software – sometimes more important. From a Human-tech perspective the only difference between management and what has traditionally been narrowly construed as technical design is the object of activity. The design of hardware and software aims to achieve an affinity with human nature at the physical and psychological levels, whereas the design of management structures aims to achieve an affinity with human nature at the organizational level.

This is an uncommon way of thinking. One of the more delightful examples to illustrate this point comes from electronic engineers' failed attempts to make it easier for people to set the time on their VCRs, an irony that probably would have made Charlie Chaplin smile.[7]

WHAT TIME IS IT, REALLY? HIGH-TECH VCR CLOCKS

Some VCR designers thought that those annoying flashing 12:00s that plague so many living rooms could be overcome by eliminating human involvement altogether. Accordingly, they designed a clock that would be automatically set by a timing signal that accompanied the transmission of a TV program. The idea seemed brilliantly logical: if the TV owner didn't have to set the time, then the potential for human error would be completely eliminated. People are fallible; they can and do drop the ball, but fancy electronic components can be counted on to do what's needed faithfully and reliably. Mechanistic innovation at its finest. Right?

Well, no. Things turned out to be a bit more complicated than that, and the effect of this fancy technology surprised not just the VCR owners, but even the engineers who designed the supposedly foolproof VCR clocks. The problem began in the San Jose, California, area, when thousands of people noticed that the self-setting clocks on their smart VCRs were running twenty-four minutes fast. In other parts of the United States, clocks were off by one, two, or even three hours. From a Mechanistic perspective, it was a paradox: even though the VCR owners didn't have to do anything, errors still occurred. It looked as if a mischievous invisible hand was messing with the VCR clocks. Weird.

To understand why this happened, let me briefly review the process by which these "autoclocks" work. An electronic circuit scans the TV channels in increasing order, beginning with channel 2. As it scans, the circuit is searching for a digital timing signal emitted by most Public Broadcasting Service (PBS) television affiliates in the United States. Once such a signal is found, the autoclock stops at that channel, and uses that timing signal to set the time on the clock. The VCR owner doesn't have to do anything; the process is entirely automated.

So why were self-setting VCR clocks around the country showing the incorrect time? It turned out that there were several different reasons – all very Chaplinesque. In one case, a PBS station in San Jose hadn't had a chief engineer on duty for a year. During that period, the timing signal had changed so that it was twenty-four minutes fast. The newly hired chief engineer didn't even know that this timing signal existed, much less that it had to be monitored regularly – there was no procedure or documentation in place at that PBS station that would have clued him in that it was part of his job. But in response to complaints from some VCR owners, he eventually consulted other technical experts, located the relevant piece of equipment, synchronized the timing signal, and thousands of VCR clocks in the area began showing the correct time. Their users must have been pleased, but I'd be willing to bet they were also mystified.

The impervious autoclocks failed to display the correct time for a second reason. The Fox TV network had started including a timing signal on the feeds being broadcast from its Los Angeles station. This signal was of the same type as that used by PBS stations, so it could be recognized by VCRs with the autoclock feature. Several Fox affiliates across the country failed to modify the signal to match their local time zone, so the Los Angeles timing signal was being broadcast in several locations across the nation. Furthermore, in many places, Fox is shown on a lower-numbered channel than PBS, causing the self-setting VCRs to detect and lock onto the Fox signal and use it to set the clock, which resulted in an incorrect display of the local time.

Imagine a Fox affiliate on the East Coast broadcasting on channel 5 and a local PBS station broadcasting on channel 13. The VCR autoclock would begin searching at channel 2, find a timing signal on channel 5, and stop its search; that signal would then set the clock to the correct time – in Los Angeles. The auto-clock would never get to the PBS station on channel 13 where the correct East Coast timing signal was being broadcast. As a result, a VCR owner on the East Coast would have their VCR clock automatically (and mysteriously) set to West Coast time, causing a discrepancy of three hours. Similar problems were experienced by VCR users in the Mountain and Central time zones, causing discrepancies of one and two hours, respectively. It took engineers about a year to uncover the convoluted and intricate set of infrastructure relationships that led to the errors.

After this phenomenon was reported in the *IEEE Spectrum* magazine, many readers wrote in to complain of their frustrating experiences with self-setting VCRs. Since VCRs are the bane of so many people, here – for your amusement – are a few of the stories.

- One reader reported that the time on his VCR would change during the day in a seemingly random way. Sometimes, the time was accurate. Then, it would be wrong. Then it would go back to being accurate – as if it was playing tricks. The cause was probably programs taped for later broadcast complete with the timing signal. When they were aired they sent out their original signals, which were now quite inaccurate, and the VCR autoclock – being technically reliable – dutifully changed the time setting every time it received a new signal. After the program ended, an accurate timing signal was broadcast again, and the clock returned to the proper time. Imagine what this would look like from the viewpoint of the bewildered VCR owner; you'd probably think your "advanced" VCR was possessed.

- A man living on the East Coast reported that his VCR was two hours fast. He fixed the problem by fooling the VCR into thinking that he lived in the Mountain time zone, thereby cancelling out the two-hour discrepancy. A few weeks later, the problem with the timing signal at his local station was corrected, so the VCR clock went back to being two hours fast. The person fixed the problem by resetting the time back to the Eastern time zone. (Dogged human adaptability trumps Mechanistic "sophistication.")

- Another man, who was an engineer himself, bought a new VCR and a new TV at the same time, both with self-setting clocks. However, the two autoclocks weren't designed to work in the same way, so they each displayed a different time. The engineer tried several means of getting the two clocks to agree; he even read the owner's manuals for both devices cover to cover. All attempts at fixing the problem were futile. (Human adaptability has its limits.)

Several people who wrote in to describe their frustrating experiences said they had disabled their VCR autoclocks and had gone back to setting the time manually.

This tale of the crazy clocks exposes an important weakness: engineers tend not to view organizational factors as an essential part of technological system design. Nobody seems to have thought about the organizational infrastructure – the tangled web of relationships – needed for the clocks to continue to work effectively. Certainly not the national television networks, local television station affiliates, and chief engineers at local PBS stations across the country – how would they all magically know what to do to make the autoclocks work? After all, they weren't the ones who designed the new VCRs. For the system as a whole to succeed, an organizational infrastructure would have had to be deliberately designed (alongside the autoclocks) so that all the relevant organizational entities could play their part in keeping time. But the one-eyed disciples of the Mechanistic view were fixated solely on the innards of the autoclocks – and, indeed, there was absolutely nothing wrong with the electronic circuits.

Getting the technical details right is obviously necessary, but it's not enough. Management also matters; it's an intrinsic part of system design, not some separate thing that can be dealt with after the fact by someone who doesn't understand the technical details. If this is ignored, the system as a whole simply won't work properly. In this relatively simple situation of the VCR clocks, the lesson is clear: meticulous attention was devoted to the design of the circuits, but their function also depended on the operation and scheduling procedures of the television networks, and no thought at all was given to that. If complex technological systems are to fulfill the purpose they were designed for, they must also be designed to have an affinity with human nature at the organizational level. And if that lesson is essential to the success of an autoclock on a VCR, just imagine how much more important it is in a complex safety-critical system, like aviation or health care, where human lives are at stake.

LEARNING FROM EXPERIENCE BEFORE PEOPLE DIE: THE AVIATION SAFETY REPORTING SYSTEM

In early December of 1974, TWA Flight 514 was flying at night in the Washington, D.C., area, west of Dulles International Airport. As the Boeing 727 was about to land in bad weather, it hit high ground. Several people were killed. In the ensuing investigation, analysts learned that virtually the same accident had been narrowly avoided about six weeks earlier. A United

Airlines DC-8 crew, also flying at night into Dulles, had barely escaped hitting the very same high ground. The cause of the near miss was identical to that of the accident; it was, eerily, almost like a dress rehearsal. Both aircraft crews misunderstood the same type of instructions issued by air traffic controllers. The crew that experienced the near miss had reported the event to their airline, but that information didn't reach the crew of TWA 514, so they didn't know about, and couldn't learn from, the experience of their counterparts at United Airlines.

Mind you, there was a reporting system in place at the time, a system whose raison d'être was to share information of this type. But it wasn't used in this case. Why not?

The system was operated by the Federal Aviation Administration (FAA), the regulatory and enforcement organization responsible for aviation safety in the United States. The FAA was simultaneously responsible for receiving information about near misses (the reporting function), and reprimanding people and organizations when they screwed up (the regulatory and enforcement function). Did this organizational design encourage people to come forward and report their mistakes? No, it did not.

Let's take a specific hypothetical example: an airline pilot who is using a new computer system installed in the cockpit. Say that system is poorly designed, so it causes her to make a mistake that leads to a near miss. Nevertheless, she is able to use her ingenuity to avoid an accident, saving her own life and those of her crew and passengers. The pilot knows why she made the mistake, and is concerned that someone else might do the same thing. She also suspects that the computer system could be redesigned to make those kinds of mistakes much less likely to happen. She wants to tell the aircraft manufacturers, so that they can fix the problem. She also wants to tell all her fellow pilots, so they too can learn from her mistake and avoid the same difficulty, which might lead once again to a near miss – or a catastrophe. In short, the pilot wants to do her part to improve aviation safety. So how can she – a conscientious professional – do that? Well, prior to 1974 she would have had to report her mistake to the FAA, the same organization that was responsible for regulating and enforcing the aviation industry. But by reporting her near miss to "the feds," the pilot ran the risk of being disciplined, fined, castigated, perhaps even sued or fired – not much of an incentive for trying to do your bit to improve aviation safety. It would not be surprising if she decided to keep quiet. No one need ever know it had happened at all.

Having the same organization responsible for these two functions was a recipe for disaster because it ignored what we know about human nature. People don't tend to come forward with information if an organization "shoots the messenger." So pilots weren't very keen on sending in reports. And who could blame them? To ask any of us to volunteer information under such conditions goes against the grain. Given the mismatch between the design of the organization and human nature, the old FAA reporting system failed to provide a basis for learning from experience, and the results were deadly. The fatal 727 accident of TWA 514 outside Washington, D.C., was just one tragic consequence.

After that fatal crash, the FAA asked NASA to step in and do something about this endemic problem. Charles Billings was to play a huge role in what followed. With his grey hair, his grandfatherly eyeglasses and his confident, clear-spoken manner, Charles Billings gives an impression of great wisdom, of having an enormous amount of invaluable real-life experience to share. In Billings's case, appearances do not deceive. He has enjoyed a highly distinguished career, and has many of the "blue ribbons" that are the hallmark of a respected academic, with several degrees to his name. Until shortly before the crash of TWA 514, he was on the faculties of both medicine and engineering at Ohio State University, where he directed the training program in aerospace medicine and was involved in teaching and research in preventive medicine and aviation. He is also a fellow and past president of the Aerospace Medical Association, a member of the International Academy of Aviation and Space Medicine, a fellow of the Royal Aeronautical Society, and an associate fellow of the American Institute of Aeronautics and Astronautics. But Billings is not a typical academic. He doesn't have his head in the clouds, and he has travelled far beyond the ivory towers of academia. In 1973, he joined NASA and remained there until 1992, retiring as chief scientist of the NASA Ames Research Center. And after the crash of TWA 514, he played a central role in setting up the Aviation Safety Reporting System (ASRS), a voluntary incident-reporting system that collects and disseminates information across a wide range of people and organizations.[8]

The goal of the ASRS is to improve aviation safety by learning from near misses (which are referred to as "incidents" in the industry) *before* accidents and deaths occur. ("Accidents" are investigated in a separate, complementary process run by the National Transportation Safety Board.) The system begins in the real world where the potential hazards associated with flying are felt in a concrete and sometimes terrifyingly vivid way by front-line workers. Aviation professionals who experience near misses – usually pilots – voluntarily write a narrative report to describe the incident, why they think it happened, and perhaps even what might be done to avoid similar problems in the future. That report is then sent to the ASRS at NASA. The initial report has an identification strip so that the analyst who looks at it can later contact the person who wrote the report to ask any clarifying questions.

The report is analyzed by experts, usually recently retired pilots who are intimately familiar with the aviation industry; their job is to index the information so that it can be organized into a database. The fact that experts are involved at this stage is critical: without their extensive technical knowledge of aviation the reports could not be indexed in a way that would be useful to others later. MBA grads wouldn't be able to do the job. After the analysis is conducted, the identification strip is removed, and the report is "de-identified" by removing all the potential identifying features. This makes the reports useless for civil legal sanctions.

The database of indexed reports is then made available to researchers in government, universities and industry—anyone who has an interest in improving aviation safety. Researchers can look for patterns and trends. For example, when new technology is introduced into the cockpit, crews sometimes experience a new set of difficulties that can weaken the people-technology fit at the physical, psychological, or team levels and thereby threaten lives. By following such trends in the ASRS database, researchers can detect problems and implement safeguards before accidents and fatalities occur.

The knowledge obtained from researching the database is then published or distributed to others in the aviation industry, so that the industry as a whole – not just the person who experienced the near miss – can learn from the past. A monthly newsletter, *Callback,* provides prompt and highly visible feedback regarding recent patterns and trends, allowing airplane designers to make changes to reduce the likelihood of problems recurring. At the same time, this feedback also motivates people to come forward and share their experiences. Those who voluntarily report near misses see that somebody is listening to, and acting on, what they're saying – their report wasn't just filed away in an archive, never to be read again. People can feel good about the fact that they're doing their part to improve aviation safety. They know they've contributed to saving lives. Charles Billings explains the impact this feedback has had on the willingness of people to write and send in reports:

> People write pages of descriptions, they send tapes, they volunteer to come to our offices. They make clear in a number of ways that they want us to understand, in all its rich detail, the

complexity of the incident in which they had been involved and why they had already reported by mail. One indication of the sincerity and dedication of these people is that some have had such close interaction that they have become personal friends. There is no question about the motivation of the pilot community in general with respect to safety issues, none whatever. The reports are not grudging acknowledgement or pro forma filings but rather quite rich and human descriptions of troubling, often frightening events.[9]

It's quite clear that organizational feedback is an essential part of the system's success. The ASRS is all about learning from experience, and that can't happen unless the lessons learned by individuals are shared widely and promptly throughout the whole industry.

It may sound as if the creation of the ASRS was easy, but it was not. Designing an effective new organizational structure to enable an entire safety-critical industry to learn from experience, foresee and prevent accidents and prevent needless deaths was a huge challenge. It took someone with Charles Billings's dedication and insight into human nature to achieve it.

It's worth taking a look at the basic principles on which the ASRS was set up, because they're relevant building blocks for organizational reporting systems across a broad range of industries, and even in business. The ASRS was created to systematically address the limitations of the old reporting system by relying on five basic principles. First, reporting is *voluntary*. No one has to report, but anyone can. The system is open to any person who has information that is relevant to aviation safety. This ensures that the system casts as wide a net as possible in its search for information that can be used to learn from experience. Secondly, it's *confidential;* it protects the identities of the people who decide to report, as well as those who might be described in the report. This encourages people to contribute because they can share their experiences without sticking their necks out in public. Thirdly, it's *non-punitive;* the FAA provides limited immunity from sanctions to the people writing the reports – which also encourages reporting. In fact, in some cases pilots will send in a report specifically to ensure they won't be subject to sanctions. The goal is to find out *what's* to blame, not *who's* to blame. The ASRS also benefits from being *objective.* NASA was chosen to operate the ASRS because it's a respected, authoritative and disinterested third party. If these conditions weren't satisfied, the ASRS would be open to accusations of bias, or even worse, its findings could be dismissed for lack of expertise. Either way, people would be less likely to take the time and effort to file incident reports. Finally, the ASRS couldn't function effectively if it weren't *independent.* The FAA is still responsible for aviation regulation and enforcement, but NASA is entirely responsible for incident reporting. Thus, the ASRS is operated completely outside the FAA, and protected from political considerations that could wind up corrupting the integrity of the process. Independence ensures a firewall between conflicting functions by assigning them to different organizations, removing a fatal problem of the old reporting system, in which people with valuable information might not report at all for fear of giving regulators a stick to beat them with. These five principles show that successful incident reporting systems for learning from experience in complex technological systems can be designed to fit human behaviour and attitudes.

As of 2001, the ASRS contained over 500,000 narrative reports in its database. It costs about $2 million annually to operate. Each year about 35,000 to 45,000 incidents are reported, an average of about a hundred every day of the year. Despite the enormous number of reports received, there has never been a breach in the confidentiality of the system. Since the ASRS was first introduced in 1976, nobody has ever been identified, blamed, disciplined, sued, fired, prosecuted or generally hung

out to dry for reporting an incident. The overwhelming majority of the reports obtained are related to issues of "bad fit"
between people and technology. The insights gathered from the
database have stimulated many changes to the aviation industry,
repeatedly improving the fit each time it threatens to weaken.

The ASRS is a Human-tech success story with an outstanding international reputation. Its very existence sends a loud, clear
message to everyone in aviation – safety matters. Few people
know it, but all of us who have been on an airplane since 1976
owe Charles Billings a debt of gratitude for making our journeys
safer than they would otherwise have been. Now that the ASRS
has had such a big impact on commercial aviation, it's being
adopted as a model for improving the fit between people and
complex technological systems in other sectors as well.

The ASRS is an organizational innovation, not a strictly
technical innovation in the traditional sense. Although it was
designed, tested and is now in use it's not a widget but a concept – a detailed, rationally and imaginatively thought-out
concept. As such, it falls completely outside the limited
Cyclopean Mechanistic world view. That psychologically blink-
ered, narrow perspective assumes that the way to prevent fatal
crashes is through hardware or software innovations, such as
more reliable mechanical components or more sophisticated
computer algorithms. These strictly technical innovations do
have their place and have indeed contributed to improvements
in aviation safety. But the enormous success enjoyed by the
ASRS shows that many lives can also be saved by deliberately
creating a system of harmonious, interconnected relationships
that allow people to learn from experience, while accommodat-
ing the difficulties of how people actually behave at an organizational level. Once again, it's all about people.

But as I said before, this is an uncommon perspective. Most
other safety-critical sectors haven't followed aviation's lead in
taking the Human-tech Revolution to the organizational level.

For example, there is no health-care equivalent to the ASRS anywhere in the world (although steps are being taken in that direction, as I will show later). Having come this far with me, you probably won't be surprised to learn that many of the sectors that have failed to institute Human-tech principles have unintentionally left a calamitous wake of dead bodies in their path.

THE VICIOUS CYCLE OF DEATH: VINCRISTINE ERRORS

In March 2001, a Canadian investigative TV news program called *W-5* aired a story about medical error, entitled "The Hidden Epidemic." Three children, each from a different province in Canada, were featured. Although the children grew up many thousands of kilometres from each other and never met, their lives became intimately connected by a truly tragic set of events.

Ryan Bishop of Thunder Bay, in northern Ontario, was diagnosed with leukemia when he was five years old. He began his chemotherapy treatment in March 1986. Just before Christmas, in December of 1989, Ryan received his second-last treatment. His leukemia had been in remission for two years, but the doctors wanted to complete the treatment, just to be on the safe side. Afterwards, on his way home, Ryan began to feel ill. He hadn't experienced any major side effects from the chemotherapy before. His family did what they could to make him feel better, but the next day Ryan felt worse, so he was taken to the emergency ward of a nearby hospital. The doctors told Ryan's parents to keep a close eye on him and sent him home. The following day, Ryan's condition was alarming and his parents took him back to the hospital. After some investigation and consultation, the medical staff realized what had happened.

Ryan had been given vincristine, a chemotherapy drug that's used to treat childhood leukemia. Vincristine is supposed to be administered only by intravenous (IV) injection. As part of the treatment, Ryan had been given another drug at the same time. That drug, methotrexate, was administered intrathecally (IT) by a spinal injection. The doctor treating Ryan mistakenly administered both drugs by IT. When vincristine is delivered in this way, it causes fatal damage to the patient's nervous system. There is no known antidote. Ryan Bishop died on December 27, 1989. He was eight years old.

An inquest was held and a coroner's jury developed a set of relevant and sensible recommendations to ensure that this type of error would never happen again (if they were implemented, of course) – recommendations that reflected both physical and psychological human factors. For example, the jury recommended that drug packages should be clearly labelled and vividly colour-coded to indicate whether they should be given by IV or IT injection, making it less likely that anyone would confuse them. The syringes should also be labelled in a salient and clear way, adding another source of crucial information. In addition, hospitals should put procedures in place so that IV and IT drugs would never be located in the same room at the same time, thereby making fatal errors less likely. Ryan's death made the front page in the local newspaper. The results of the coroner's inquest were widely publicized in Thunder Bay (but not more broadly) so that others would know what had happened and what could be done to prevent another such tragedy.

Ryan Bishop's family was distraught with grief after the loss of their young son, but they found some comfort in the recommendations that had been made by the coroner's jury. Ryan's mother told *W-5* that the family had decided not to pursue legal action because: "We all make mistakes. We just wanted to make sure it never happened again."

But could this goal be achieved merely by improving (as the recommendations attempted to do) the bad fit between people and technology – the drug packaging, the clear labelling, the hospital procedures – at the physical and psychological levels, the bottom two rungs of the Human-tech ladder? Unfortunately, it could not.

Courtney Braund of Yarmouth, Nova Scotia, was diagnosed with leukemia in 1990 when she was just two years old. Like Ryan Bishop, she was treated with methotrexate and vincristine. After two years of treatment, her cancer had gone into remission. In April 1992, Courtney received her last treatment. Her doctor was busy that day, so a different doctor administered the drugs. Like Ryan's doctor, he mistakenly delivered both drugs by IT injection, rather than delivering vincristine by IV injection. Over the next few days, Courtney's parents noticed that she wasn't her usual self. When she didn't get better, they took her to the hospital. It was only at that point that the irreversible error was detected. Courtney Braund died on April 30, 1992. She was four years old. After her death, test results found no cancer cells in her body. The chemotherapy treatment had worked. She had already been cured of leukemia.

Two years later, in September 1994, four-year-old Kristine Walker of Sidney, British Columbia, was diagnosed with leukemia. She completed two years of chemotherapy treatment and her cancer went into remission. A year later, the leukemia showed signs of returning so a new round of treatments was planned. Like Ryan Bishop and Courtney Braund, she was treated with methotrexate and vincristine. On May 21, 1997, Kristine began her second round of chemotherapy. Again, a doctor mistakenly delivered vincristine by IT injection instead of by IV injection. Again, the doctor and his hospital were unaware that this tragic, avoidable mistake had been made before and preventative steps recommended by investigators. In this case, the error was detected right after it was made. The doctor sought advice. A number of medical procedures were performed, but to no

avail. Over the next few weeks, Kristine's condition degenerated
to the point where she had massive neurological damage. She
died on June 3, 1997. She was seven years old.

Kristine's parents sued the hospital, not for money, but to
make sure that changes were made so that the same thing
would never happen again to another child. They believed that
blaming people wouldn't make things better, but that the health-
care system itself needed to be designed differently to prevent
this type of error. A number of fundamental changes were put
in place at that hospital, several of them similar to the recommendations that had been made by the coroner's jury in the case
of Ryan Bishop about eight years earlier.

The vincristine problem isn't, of course, just a Canadian one.
In the United Kingdom, there have been thirteen similar cases.
Repeatedly, those who learn of each case lament the loss of life,
express tremendous frustration, and wonder what it will take to
put an end to the sad cycle of death once and for all.

These cases suggest that the health-care industry, unlike aviation, hasn't yet found a way to learn from its mistakes by reporting and sharing information. Research backs up this impression:
only 1.1 per cent to 7.7 per cent of the medical errors that occur
ever get reported.[10] In one study, researchers at the University of
Texas found that, on average, fewer than five adverse drug reactions were reported per year in their large 900-bed hospital, but
when they looked closely at the patient records to see how
many adverse drug reactions they could detect, they found an
average of 240 per year – *50 times more* than people had been
reporting, leading to a reporting rate of a mere 2 per cent.[11] This
is precisely what used to happen in aviation back in 1974. When
the FAA was responsible for receiving information about near
misses *and* reprimanding people and organizations, very few
people came forward with information.

The under-reporting of medical errors comes at a tremendous
cost to society. If we view error as an opportunity to improve

patient safety, then a reporting rate of 1.1 per cent to 7.7 per cent means that for every incident report that we receive there are somewhere between 12 and 89 other events that occur but aren't reported. Almost all the opportunities for reducing medical error are going down the tubes, which results in a catastrophic failure to learn from experience.

Why has the health care sector had such a difficult time creating learning organizations? The simple answer is fear. Doctors and nurses are reluctant to volunteer information about problem areas that are in need of improvement because they're scared of being punished for doing so. But why would such fear exist? I will discuss two big contributors: legal liability and the medical culture of personal accountability.

SEEKING SCAPEGOATS VS. SAVING LIVES (PART I): THE DENVER NURSES TRIAL

Legal liability is a powerful but sometimes dysfunctional force that puts a substantial obstacle in the way of those who want a Human-tech Revolution in health care. Legislators don't intend to harm patients, but laws can nevertheless make it difficult for organizations to benefit from experience. American product liability law provides a good illustration of this point.[12] The law states that if a company improves the design of its product in response to an accident, that design change can't be used by a plaintiff as evidence of prior negligence in a lawsuit against the company. The rationale is that if companies run the risk of being punished for making improvements, they won't make changes, refusing to admit that the product is deficient. The result will be more accidents that could have been avoided. So the law says that improving a product can't be used to prove the manufacturer was negligent.

However, the law is far from perfect. Evidence of a design change after an accident is "discoverable," which means that the plaintiff's lawyers can find out about it. And they can put it in front of a jury to suggest the manufacturer knew that a design improvement *was* technically feasible. When this occurs, a judge can instruct a jury to use that evidence only to determine whether an improvement existed, not to infer negligence. But as legal experts have pointed out, in practice such instructions may have little effect when you have real human beings made of flesh and blood sitting in a jury box rather than the "perfectly rational" humans assumed by the law; when jurors hear evidence that a product improvement was technically feasible, but was only introduced after an accident, they may still convict the company – regardless of how or why such evidence was introduced, or what special instructions were given – because they're likely to believe the accident was preventable.[13] So a company that wants to make a design change is faced with the prospect of having that action used as evidence against it, and paying big bucks in product liability lawsuits. This is not a good way to encourage design changes that will save lives.

But that's not the only way that legal liability can perpetuate lethal medical errors. In fact, the justice system's counterproductive effects on patient safety are so numerous that it took the legal scholar Professor Brian Liang thirty-two dense pages to describe them (and he didn't cover the whole territory) in a peer-reviewed scientific article, entitled "Error in Medicine: Legal Impediments to U.S. Reform" – an outstanding analysis of the legal "invisible hands" that unintentionally continue to harm patients.[14] Rather than bury you in legal jargon and case law, I'll use a case study to illustrate the general point.

The widely publicized Denver nurses' trial shows how the justice system can stand in the way of health-care reform.[15] The basic facts of the case are not in dispute. On October 15, 1996, a thirty-two-year-old mother of three went to the hospital to deliver a baby. She had a history of syphilis dating back to the early 1980s, but there was no record of the treatment she had received. During her pregnancy, lab tests for syphilis had come back positive. Her doctor hadn't provided any treatment during the pregnancy. At 9:59 A.M., the mother gave birth to a baby boy. The doctors ordered a number of tests to see if the baby also had syphilis.

Since it would take a while for the test results to become available, the doctors began the baby's treatment right away. A prescription was written for a particular type of penicillin. In preparing the two syringes, the pharmacist made a mistake. She filled the prescription with ten times the intended dose of penicillin – an error that went undetected at the time.

Three nurses were involved in giving penicillin to the baby. The drug was supposed to be given by intramuscular (IM) injection, but one of the nurses suggested that perhaps the drug should be given intravenously (IV) instead. None of the nurses had ever given this type of penicillin before. They looked up the drug in a reference book, and found that penicillin could indeed be given IV. However, the nurses didn't know that there are two types of penicillin: aqueous and viscous. The type that had been prepared for the baby was viscous. The aqueous type can be given by IV injection, but the viscous type can only be given safely by IM injection. The reference book they consulted didn't mention this. So the nurses decided to administer the drug by IV instead of by IM injection, as they should have done.

At 2:18 P.M. on October 16, the one-day-old baby boy received his first injection. One of the nurses was holding the baby's arms and legs. After the injection was started, the nurses' worst nightmare was realized – the baby became limp and experienced a heart attack. The nurses tried to resuscitate him and after fifty minutes, his heart was beating again. But he later suffered another heart attack and died at 5:04 P.M. On the same day, the test results came back. They were all negative. The baby never had syphilis.

Such an event is a tragedy for all involved – the family, the pharmacist and the nurses. The level of emotional trauma experienced by all these people is unimaginable. But what makes the Denver nurses' trial astounding is that that was just the beginning of the story. What happened next surprised almost everyone, and sent a shock wave throughout the health care community in the United States and beyond. The Adam County district attorney, Robert S. Grant, decided to indict the three nurses – but not the pharmacist or any doctor involved in the case – for criminally negligent homicide. Under such circumstances, it wouldn't be surprising to see a civil lawsuit for damages launched by the bereaved family against, say, the hospital. Indeed, financial compensation is a just remedy for patient injury and death caused by error, a point that I will address in more detail later. But for an elected district attorney to indict the three nurses under criminal law was something entirely different. His decision wasn't a simple case of seeking just compensation for a tragic loss; it was a deliberate choice to lay blame and to punish. A grand jury was convened and decided that there was enough evidence to go to trial.

Two of the nurses settled their cases out of court, but the third went to trial and was represented by the defence attorney Charles Torres. Evidently, he did a very good job of defending his client – after less than an hour of deliberation, the jury returned a verdict of not guilty.

Was the criminal prosecution justified? I was given a ringside view on the question in November 1998 when I attended a session on the Denver nurses' trial at the Annenberg Conference, which was held in sunny Rancho Mirage. The risk manager for the hospital where the tragic event occurred was there to present the facts of the case. She made no arguments in favour of either side. District Attorney Grant was there to justify his decision to charge the nurses with criminally negligent homicide. The defence attorney, Torres, was there to explain why his client wasn't at fault, and why he thought Grant had decided to demonize the three nurses. Grant and Torres had been bitter adversaries in a hard-fought legal battle. Now they were in a packed auditorium with an audience of three hundred attentive spectators keen on improving patient safety. The tension in the air was palpable.

District Attorney Grant's primary concern, he explained, was accountability. A one-day-old baby had died; somebody had to be held accountable in order for justice to be achieved. Prosecution was necessary as a deterrent to prevent such mistakes in the future. The logic of this argument is not only understandable, but familiar. We all feel that if wrong has been done, those responsible must be held to account. From this perspective, the prosecution is clearly justifiable. In fact, the argument is *so* compelling that there doesn't seem to be any room for an alternative. What else can we do: let the culprits off scot-free?

From a Human-tech perspective, the question is just the beginning of the discussion, and there *is* more than one alternative. If the Denver nurses had been wilfully negligent, then yes, seeking accountability via criminal charges would indeed have been appropriate. Bad apples should be dealt with accordingly. But if the nurses made an honest mistake, prosecuting them wouldn't do any good because they were already trying to be conscientious and diligent, although financial damages would be a proper judgment to the bereaved family.

So what camp did the Denver nurses fall into: bad apples, good people who made a terrible mistake, or something in between?

The newborn baby had just been through a painful procedure and was crying. The penicillin came in a syringe with a 19-gauge needle – think big, really big. Rather than inflict even more pain on the baby by piercing his skin with the huge needle, the nurses checked whether the drug could be delivered IV instead of IM. So far, as Torres argued, this was not a case of bad apples. In fact,

Mike Cohen – president of the Institute for Safe Medication Practices, and one of the expert witnesses who testified at the trial – later told me that there were "over 50 systemic failures, almost all of which were not in the nurses' control, all merged at the same moment in time to set the nurses up and involve them in this tragedy" – a quintessential *systems accident:* death by interactions, if you will.[16] Torres set out a number of facts to support the same conclusion (this is just a partial list):

- The nurses weren't accustomed to dealing with the drug because penicillin isn't common in a neonatal ward. One-day-old babies don't usually have syphilis.
- The reference information the nurses properly consulted was incomplete because it didn't specifically mention the kind of penicillin that had been prepared. From the nurses' perspective, they followed the advice in the book by giving the drug IV.
- The pharmacist made a tenfold mistake in preparing the penicillin that contributed to the death, yet she wasn't charged,[17] while the nurses were. Why? District Attorney Grant responded to this by saying that the pharmacist made a "medication error" not a "human error," a bizarre distinction that baffles me to this day.
- A nursing expert testifying on behalf of the prosecution made the same math error that she testified the nurses had made. In fact, she made the error about eight or nine times, until a grand juror brought the error to the expert's attention, at which point she still couldn't figure out what the error was. The expert handed a calculator to the grand juror who then performed the calculation to demonstrate the mistake. At this point, the district attorney had to tell the grand juror to give the calculator back because the witness was supposed to be the expert, not the grand juror. Afterwards, the expert witness said something to the effect of "We're all human. Nurses make mistakes."

- A pharmacist also gave expert witness testimony to the grand jury on behalf of the prosecution. Later, he admitted to having made six mistakes. During the trial, the prosecution replaced him with a different expert witness. Ironically, that new pharmacist wound up making so many points in favour of the defence that the prosecuting attorney wanted to impeach the expert. The judge had to stop the prosecutor from doing that and remind him that the pharmacist had been called by the prosecution as their own witness.

- The attending physician testified that the nurses should have asked her if the route of drug administration could be changed from IM to IV, and that she wouldn't have allowed such a change. However, several nurses then testified that, although consulting with the attending physician is the official rule, in practice, doctors tell nurses to make such decisions on their own – "don't bother me unless there's a significant problem" was the routine practice at the hospital (and many others as well).

- According to Torres, "interviews of jurors afterwards indicated that they felt the prosecution witnesses had been untruthful and believed the district attorney had attempted to present false information to the jury."

The evidence was so compelling that there was no grey area in this case. The jury concluded that the nurses were not bad apples, that they were concerned for the welfare of the newborn child, but that they faced a situation where error was likely to occur. And it took them less than an hour to reach that conclusion.

Instead of charging nurses with homicide, a more constructive remedy would have been to understand why the tragic error occurred. If we believe that human life is precious – and presumably we all do – then, difficult though it may be, we should resist the impulse to rush to judgment. We should also examine the context in which the tragedy occurred, and ask a few more questions. Will prosecution under criminal law improve patient

safety, or prevent the death of *more* one-day-old babies in similar circumstances? Will it mean that no parents will ever again
have to live with the lifelong grief of having had to bury their
son or daughter after a preventable accident? The rationale
behind legal prosecution is that an individual must be at fault,
and this may be true. But the design of the system may be at
fault. If we focus on finding and blaming scapegoats, we may not
investigate this option, and we won't be able to identify and
address the systemic factors that are causing well-intentioned
and properly qualified people to make mistakes, and then
redesign the system to improve patient safety. We can't undo
the past, but we can learn from it by looking to the future.

Mike Cohen summarized the fundamental lesson of the Denver
nurses' trial by emphasizing the profound mismatch the case
demonstrated between the two parts of the equation: the design
of the technological system in which the nurses worked, on the
one hand, and human nature on the other hand. We must learn
from experience, he said, by designing organizations that have
an affinity with human nature: "We must look beyond blaming
individuals and focus on the multiple underlying system failures
which shape individual behaviour and create the conditions
under which medication errors occur. And by their verdict, this
is the lesson that the Denver jury wants us to learn."[18]

SEEKING SCAPEGOATS VS. SAVING LIVES (PART II):
THE "BLAME AND SHAME" CULTURE

It's not just the threat of legal liability that causes doctors and
nurses to be afraid of admitting to error. As I mentioned before,
the health-care sector has traditionally held the view that the
individual medical professional must accept the prime responsibility for errors, regardless of how systems are designed; doctors

are supposed to have superhuman wisdom and strength, even if they haven't slept in thirty hours and are trying to read a prescription order that looks like chicken-scratch at 4 A.M. From this perspective, the logical thing to do when errors occur is to focus on the person who is immediately responsible. In health care, people are blamed, told to be more attentive, given remedial training, disciplined and sometimes fired. This isn't a secret, and it's not a matter of opinion; it's an unassailable fact borne out by research. As I will show below, things are slowly starting to change, but the traditional "blame and shame" approach to dealing with medical error is recognized by everyone who works in the sector, and it has been openly discussed by doctors and nurses themselves. I quoted Lucian Leape on this score, in chapter 4, but he's far from the only medical professional to have discussed the issue.[19] And as with legal liability, punitive or remedial measures may be appropriate in the case of bad apples or real negligence, but in cases where people are already doing their best, and are already well trained, "blame and shame" tactics are irrelevant because they don't get at the root cause of the problem.[20]

Actually, I take that back. Blaming and shaming aren't irrelevant; they're *worse* than irrelevant because they can contribute to errors. Looking for an individual to blame when something goes wrong, regardless of the conditions that person is working under, is tantamount to creating a gigantic "invisible hand" that's ready at all times to point an accusing finger at anyone who makes a mistake. Such a design can threaten patient safety by making it less likely that competent medical professionals will come forward to report and identify deficiencies in the healthcare system.

After the thirteenth vincristine error in the U.K., Don Berwick, president and CEO of the Boston-based Institute for Healthcare Improvement, noted that people were outraged, and asked, "How could this happen – again?" His answer is thought-provoking:

"The answer is surprisingly mundane. It is this: we are human, and humans err. Despite outrage, despite grief, despite experience, despite our best efforts, despite our deepest wishes, we are born fallible and will remain so."[21] Targeting the errors of individuals as the sole, or even primary, cause of medical accidents has not improved patient safety and never will – but it is the default reaction, over and over again. The usual suspects were trotted out and made examples of after the latest vincristine error in 2000 in the U.K.: doctors were suspended; an investigation was started; doctors were exhorted to try harder. The fact that deaths from vincristine – and from other medical errors – continued to occur in the United Kingdom, as well as in Canada and other countries, shows that this approach simply doesn't work. As Berwick says: "Expecting perfection in human action, or simply telling doctors and nurses to 'try harder' – not to kill their patients by mistake – has nothing at all to do with our eventual success."[22] Most health-care professionals are already trying hard. Except for the rare bad apple, nobody wants to see patients injured or killed. To expect health-care providers to be perfect is equivalent to asking them to stop acting like human beings. It is the design of the health-care system itself that must undergo change and development.

The base human temptation to always find a culprit is so strong that it may be difficult for some readers to believe that blaming and punishing health-care professionals who make mistakes is counterproductive. Don Berwick says mistakes are inevitable and should not be punished, and he is one of the most respected patient safety experts in the world, but it's easy to dismiss a single voice as a fringe view. So here's some more compelling evidence. The case for the pervasive "blame and shame" culture actually making things worse is supported by the evidence of aviation pre-ASRS compared to post-ASRS; it's a finding that has been reported by world-renowned researchers in peer-reviewed articles published in the best medical journals in the world, and echoed by the U.S. IOM – the most prestigious, authoritative and impartial group of medical professionals in the United States.[23] Just think about the kind of scrutiny that this claim was subjected to before it was printed in these prestigious publications, publications that are edited and reviewed by medical professionals themselves.

The case for putting all of the responsibility on the individual fails, in part, because it overlooks the deterrents that already exist to avoid medical error – the incredible anguish and remorse that many doctors feel when they harm a patient.[24] Dr. David Hilfiker vividly described these deterrents in a courageous 1984 article in the *New England Journal of Medicine*, in which he publicly confessed to some of the medical mistakes that he had made in his rural family practice.[25] In one case, Hilfiker agreed to help deliver the baby of his friends, Barb and Russ. The pregnancy seemed to be going well, except for one thing; Barb's pregnancy test was negative. At first, Hilfiker attributed the unexpected result to the early stage of the pregnancy. He told Barb to come back the following week to perform another test, one that would surely confirm what was already visible to the naked eye – Barb was expecting. But the second test came back negative as well. Hilfiker considered the options; he could order an ultrasound, but it would have been expensive and would have required his friends to drive 175 kilometres. Instead, he waited a month and ordered a third test. By this time, Barb still hadn't had a menstrual period and her uterus continued to be enlarged, but the third test also came back negative.

Hilfiker decided there was only one explanation that could possibly account for all of these seemingly contradictory facts. So it was with great regret that he told Barb:

> You were probably pregnant, but the baby appears to have died some weeks ago . . . Unfortunately, you didn't have the miscarriage to get rid of the dead tissue from the baby and placenta. If a

miscarriage does not occur within a few weeks, I'd recommend
a reexamination, another pregnancy test, and if nothing shows up,
a dilation and a curettage to clean out the uterus.[26]

When Barb returned after two weeks, she had not had a miscarriage and the fourth pregnancy test was negative, so the surgical procedure was called for two days later.

Only the unhappy words of Hilfiker can do justice to what
happened during the operation:

> I examine Barb's pelvis. To my hands, the uterus now seems
> bigger than it had two days previously, but since all of the preg-
> nancy tests were negative, the uterus couldn't have grown. I
> continue the operation. . . . The body parts I remove are much
> larger than I expected, considering when the fetus died, and
> they are not the decomposing tissue I'd anticipated. These are
> body parts that were recently alive! I suppress the rising panic
> in my body and try to complete the procedure. . . . Despite
> reassurances from the pathologist that it is statistically 'impos-
> sible' for four consecutive pregnancy tests to be negative dur-
> ing a viable pregnancy, the horrifying awareness is growing
> that I have probably aborted Barb's living child . . . The pathol-
> ogy report confirms my worst fears: I have aborted a living
> fetus at about 13 weeks of age. No explanation can be found for
> the negative pregnancy tests. My consultation with Barb and
> Russ later in the week is one of the hardest things I have ever
> done. . . . nothing can obscure the hard reality: I have killed
> their baby.[27]

Hilfiker went on to describe the guilt and anger he felt. He
replayed all the events in his mind to see what he could have
done differently. He blamed himself. Again, his own words
explain the devastating emotional impact that a doctor can
experience after making a mistake that harms a patient:

Everyone, of course, makes mistakes, and no one enjoys the consequences. But the potential consequences of our medical mistakes are so overwhelming that it is almost impossible for practicing physicians to deal with their errors in a psychologically healthy fashion. Most people–doctors and patients alike–harbor deep within themselves the expectation that the physician will be perfect. No one seems prepared to accept the simple fact of life that physicians, like anyone else, will make mistakes.[28]

Hilfiker's brave public confession shows the personal cost of medical error to the people who commit the errors.

His feelings are consistent with the experiences of Donnalee Braund. About a year and a half after her daughter, Courtney, had died from a misadministration of vincristine, Donnalee met with the doctor who had treated her daughter. She was furious and wanted to vent her frustration. However, after her meeting with the doctor, Donnalee reported to *W-5* that "he was in worse shape than I was." The doctor, who had dedicated his entire professional career to helping children, was distraught. He had been seeing a psychiatrist, but he couldn't come to grips with the mistake that he had made and the lethal effect that it had on four-year-old Courtney Braund. Eventually, Donnalee forgave the doctor for his mistake.

Once I learned to see things from this perspective, it seemed to me the health-care profession already has a very strong natural deterrent in place, in the hearts and minds of its practitioners – the raw psychological trauma vividly described by Hilfiker and witnessed by Donnalee Braund. Given this built-in "invisible hand" that already motivates doctors and nurses to avoid errors, measures like the threat of a lawsuit or suspension or exhortations don't mean much as additional constructive incentives. Don Berwick is of the same opinion:

Just "trying harder" makes no one superhuman. Exhortation does not help much, nor will suspending doctors, nor will outrage in the headlines, nor even will guilt. Suspend every doctor today who makes an error today, and the error rates . . . tomorrow will be exactly the same as today's. There is no remedy to be found in selecting heroes, nor in seeking Superman. Tomorrow, like today, we will be human. The remedy is in changing systems of work. The remedy is in design.[29]

I've discussed this issue with many people and my arguments sometimes meet with great resistance. One objection goes something like this: "So you're trying to tell me that you want to let everyone who makes a medical error off the hook, with no repercussions? That's a joke. How would you keep doctors and nurses in line without the threat of punishment?" My response is that, of course, financial compensation should be paid in cases where patients are harmed by substandard care, medical licences should be revoked if doctors are shown to be quacks, and criminal prosecution is a just remedy for doctors or nurses who deliberately harm their patients. But anyone who thinks that the only way to keep health-care providers from making mistakes is to always point a finger of "blame and shame" doesn't appreciate how powerful the desire to avoid error already is. Every single one of us makes mistakes on the job (probably every day, if you include trivial errors), and though most of us don't know what it's like to work in an environment where our actions can injure, maim or kill, it shouldn't take a huge leap of imagination to realize that, like David Hilfiker, most people wouldn't enjoy knowing that they've killed another human being – or from having to tell the dead patient's family what happened and face their shock and grief when they're given the terrible news.

Sometimes, the discussion doesn't stop there. I've also been told (usually, in a rather forceful way): "That's easy for *you* to say – forgive and forget. But what if something happened to one of *your* relatives, wouldn't you want to sue the doctor or have him prosecuted?" My answer to this is also straightforward: I *have* had a loved one harmed by a medical procedure. My grandmother was visiting Canada from Portugal during the winter when she slipped on a snow-covered sidewalk, fell and broke her leg. The doctor who took care of her decided to put her leg in a cast rather than operate on it, but he didn't set it correctly. As a result, my grandmother had to live with one leg significantly shorter than the other for about twenty years, hobbling around uncomfortably. Was my family upset? Absolutely. Did we expect better from our health-care system? Definitely. Did we sue the doctor for damages? No.

I've also been asked (usually, in an even more forceful way): "Are you trying to tell me that if you had your kid die from a medical error you wouldn't want to go for the throat and seek the blood of the people who were responsible in any way you could? How would you feel under *those* conditions?!" My honest answer is: I don't know how I would react. I can't imagine what it must be like to go through such a tragedy. My guess is that, at least initially, I would think my life wasn't worth living any more. But I *do* know how I'd *like* to be able to react – precisely the way Kristine Walker's parents did after their daughter was killed by the misadministration of vincristine. Recognizing the complexity of the circumstances, they resisted the urge to punish a culprit, and summoned up the courage and wisdom to make the following constructive and far-sighted public statement: "Our daughter gave her life for others–so that the medical system could review itself and make recommendations that would benefit all children to follow. This case was an opportunity for positive change. That is our daughter's legacy." Truly remarkable as their words are, the Walkers' reaction is not unique. Ryan Bishop's mother didn't sue her dead son's doctor, and Donnalee Braund forgave her dead daughter's doctor. I don't know if these reactions are atypical, but they demonstrate that even parents who have lost their children to medical error have managed to make the extraordinary effort to "see" the system as a whole.

But with the exception of some early signs of reform, the health-care profession continues to take the blame-and-punish approach, which creates a huge psychological roadblock to the widespread adoption of a Human-tech approach. Just as the excessive spraying of pesticides has exacerbated the problem of boll weevil in cotton, just as the overprescription of antibiotics has created new strains of drug-resistant bacteria, and just as violent use of "anti-terrorist" retaliation only begets more violence, the indiscriminate blame-and-shame culture in health care has unwittingly fanned the flames of medical error. If we value human life over vengeance, we must put the need to save lives ahead of the understandable desire for retribution. In short, we need to follow the example of aviation, and design learning organizations in health care to put an end to the cycle of preventable deaths and injuries. Fortunately, designing an organizational structure that allows its members to learn from experience, that lays emphasis on the receiving and communication of information that would help prevent medical errors, is closer to being realized than ever before.

SIMPLE BUT REVOLUTIONARY:
HUMAN-TECH REFORM IN HEALTH CARE

The lessons learned in the aviation industry are slowly being heeded by the health-care sector, and some profound changes are gradually taking hold. Ellen Tracy is one of the people who has participated in these promising beginnings. She is a registered nurse and oncology nurse manager at the Children's Hospital in Philadelphia, Pennsylvania. Tracy speaks with a passion and commitment that come from having gone through a life-altering experience.[30]

Her story begins with a decision made by the chief operating officer at her hospital. He had heard about a quality-improvement program to reduce drug administration errors, so he sent her and some of her co-workers to learn how to implement the program in their hospital.[31] He also said that he would provide her with all the support she needed to make the program work. She didn't believe him at the time, but she went to the course anyway.

At the time, the policy at Children's Hospital was to immediately discipline any nurse who made an error. The goal was to reduce human error by instilling fear. If an error was made, somebody was automatically at fault. There was no process in place to identify any underlying conditions that might have contributed to the problem.

During the program, Tracy was taught by the instructors that this traditional attitude was inappropriate because it wouldn't have the desired effect – to reduce errors. She learned about inevitable human fallibility and how feedback at the organizational level can play a critical role in identifying errors, fixing problems that cause errors, and thereby enhancing patient safety. She was told that fear isn't an effective way to combat error. A punitive system only makes people try to cover up their mistakes, removing important opportunities for improving the health care system.

Tracy wasn't sure the organizational policy that her hospital had in place really needed to be changed. But shortly after her return, something happened that started her on her own personal Human-tech Revolution. On a Saturday-night shift, a nurse made two chemotherapy errors. Fortunately, nobody was killed, but the errors were still very serious. The other nurses Tracy worked with were ready to follow the old policy. Tracey was told, "You must suspend her. . . . She must be taught a lesson, period, the end."[32] When she sat down with the nurse,

Tracy found that she too was ready to accept the consequences of the old policy. "I accept full responsibility for my practice. I'm sure that I am going to be suspended," she told Tracy. Tracy replied, "You are going to be suspended." At that point, the nurse started to cry and said, "I am fully responsible."[33]

At that critical moment, the internal conflict that Tracy had been going through rose to the surface. She was no longer comfortable following the old policy of attempting to correct errors by blaming people and instilling fear with the threat of discipline. She decided that she should at least try to identify the factors that had led the nurse to make the two errors. So she said, "You cannot leave my office until you tell me every single thing that happened from 7:00 P.M. to 7:00 A.M. [on your twelve-hour night shift]."[34]

As the guilt-ridden, sobbing nurse explained the conditions that she had had to deal with, Tracy realized the situation was more complex than it seemed. Given the number of things that were going on in the ward at the same time, and the fact that another nurse had called in sick, she was surprised that the nurse had only made two errors and not ten. She changed her mind about suspending her, and instead, began a process that forever changed the way that hospital dealt with medical error.

She called all the nurses together and informed them that in the future they wouldn't be automatically disciplined if they committed an error. They were told they should report all the things that *almost* went wrong – close calls, near misses, any place where there was room for improvement. According to Tracy, you could almost feel the sense of relief experienced by the nurses in the room. Before, they had the threat of discipline always hanging over their heads; when they made a mistake – or almost made one – they knew they would always be personally singled out and disciplined – regardless of the contributing factors. Under the new policy, they were asked to identify areas that could be improved. It increased their personal sense of fulfillment and accomplishment – now they were working together toward improving patient safety.

The effect of this simple but revolutionary change was eye-opening. At first, the result seemed alarming. The number of near misses reported by the nurses increased from two to eighteen in one month, and reached a peak of twenty-one reports two months later – a tenfold monthly increase in near-miss reports. Were Tracy's high hopes misplaced? No. Because the nurses no longer had to fear being automatically disciplined, they were now reporting *potential* problems – threats to patient safety that had always been there, but that they had been afraid to mention under the previous policy. These results are a vivid indication of the price organizations pay by always seeking scapegoats: 90 per cent of the information they now brought forward – information that was useful in preventing error and potentially saving lives – had all been suppressed and essentially wasted.

Eventually, the number of near-miss reports fell back down to about three, after many of the problems that had plagued Tracy's unit had been identified and solved. There was no longer so much to report. More important, the number of errors – not near misses, but actual mistakes that could injure or kill patients – was *reduced by 90 per cent.* These results provide an equally vivid indication of the benefits of Human-tech thinking: the new policy didn't result in mayhem, with nurses and doctors carelessly harming patients left, right and centre because they knew they wouldn't get blamed and punished; on the contrary, the new policy was responsible for a remarkable reduction in errors. A change in organizational design had taken place, generating information that allowed the hospital to learn from experience and significantly improve patient safety.

And there was satisfaction at the psychological level too. All the nurses felt better, knowing they were doing their best to help patients. They took pride in their enlightened, humane working environment and the tremendous results they had achieved. And Ellen Tracy couldn't wait to spread the word by telling others about the Human-tech Revolution in her work environment.

In many hospitals, health-care practitioners and administrators continue to hide mistakes from patients for fear of being sued. Why give the adversary a smoking gun that can then be used to incriminate you? This attitude may seem reasonable, but it overlooks a basic fact about human nature: contemptuous treatment breeds a desire for revenge. Patients are more likely to sue a hospital when they feel they're not being told the truth, that information is being withheld, and that the hospital is refusing to take responsibility for its mistakes. In short, our motivation to sue increases when our trust – as patients and families – is betrayed.

A Veterans' Administration hospital in Lexington, Kentucky, used these findings to design a radically different organizational policy for the disclosure of errors.[35] Patients' interests are the top priority. If a mistake is made, the administration informs patients or their families of the facts, and explains the countermeasures that have been implemented to prevent similar mistakes. The hospital also tells patients or their families that they can seek a financial claim against the hospital, advises them to get their own lawyer, and even goes so far as to help them fill out the claim forms! Patients or their families are financially compensated for substandard health care *without* having to engage in expensive, time-consuming and emotionally draining legal trials. Justice is achieved by full disclosure and fair compensation, rather than adversarial judicial proceedings. At the same time, patient safety is improved because the causes of the mistake are investigated, allowing the hospital staff to learn from experience.

The Kentucky hospital has been using this radical approach since 1987, because the staff and the administration both believe it's morally preferable to the traditional approach. But it had a surprising result: an unexpected *decrease* in liability payments. The hospital has one of the lowest malpractice payment costs in its class, about a tenth of those of the hospital with the highest payments. These results may seem counterintuitive; giving people information to use against you should hurt you, not help you. The resolution of the paradox is in the policy's affinity with human nature: people who feel they've been treated dishonestly are motivated to sue for malpractice claims out of a desire for revenge; taking away that fuel quenches the fire. The result is benign. All of those involved – patients, families and the hospital – are better able to deal with a distressing or tragic event in an open, just, constructive, humane and dignified way.

This policy is still the exception rather than the rule in health care as a whole. I've described it in presentations that I've given at health-care conferences, and I can see the look of disbelief in some of the audience members' eyes – the traditional risk-management policies still being used in many hospitals are so different from this policy of extreme honesty that it's hard for people to believe that any hospital is actually doing business this way. But the facts can't be denied; the policy has been so successful that it became required of all Veterans' Administrations medical facilities in the United States, beginning in the late 1990s.[36]

The only people who lose out are the malpractice lawyers, who have a smaller settlement claim from which to take their share.

As we move up the Human-tech ladder – from the physical to the psychological, and to the team and organizational levels – the task of creating systems becomes more demanding. Beyond designing products or systems to fit our bodies and our minds, after we have coordinated our activities to facilitate teamwork, there is still the huge challenge of organizing the structures and functioning of the overarching institutions that affect us all. And at these more complex levels, the power over our lives also increases – for good or ill. Organizational decisions, such as information flow, allocation of functions, reward structures and staffing decisions can have a significant impact, not just on a single individual or a single team of individuals, but on all of the individuals in an organization.

The *Challenger* disaster, the deadly accidents in aviation before 1974, and the vicious cycle of vincristine deaths in health care show the folly of ignoring management matters. The quality of all our lives is at stake. And yet, despite the demonstrated impact of management decisions on the success or failure of complex technological systems, the human aspects of the design of an organization are still not often viewed as an integral part of the design of technological systems.

That's the bad news. The good news is that the message is beginning to be heard. And we already have the knowledge to do better than we're currently doing. The stellar success of the ASRS shows that it's possible to design systems systematically to have an affinity with human nature at the organizational level. Granted, the forces of legal liability and the super-Humanistic culture of blame-and-shame pose key obstacles to the adoption of this radically different way of thinking in health care. But as society begins to see that these antiquated ways of thinking merely perpetuate fatal medical errors, that the overwhelming majority of medical-error injuries and deaths are not caused by bad apples, that we can't expect health-care providers to be superhuman, and that there are already strong incentives in place to avoid errors, then traditional currents of thought will gradually become impotent. There is far too much at stake to keep on repeating mistakes when the knowledge of how to do better already exists, and when the societal costs of failing to use that knowledge are so great and so widespread. Paying attention to the human factor saves lives.

But to realize its potential to improve the quality of all of our lives, the implementation of the Human-tech world view can't end at the organizational level. Which brings us to the fifth and highest level on the Human-tech ladder. The obstacles to improving patient safety show that *political* factors, such as laws, budgets, regulations and policy agendas, exert a strong top-down influence on the behaviour of the organizational level – an example of the interaction across levels I referred to earlier. When political factors ignore what we know about human nature, then dysfunctional ways of thinking, such as blame-and-shame, can be perpetuated, wreaking havoc and destruction on our societies. If we could figure out how to tailor political factors to have an affinity with human nature, then perhaps it would be possible to create a new set of forces that would shape technological systems to better serve human and societal needs. Is this just an idealistic pipe dream? I think not. I believe we know enough already to at least entertain the notion of Human-tech system design at the political level of human behaviour, as I hope to show you in the next two chapters.

8

Political Imperatives I:
Technology for Better or for Worse?

THE HUMAN-TECH LADDER

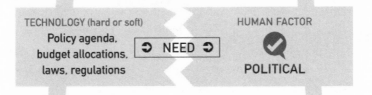

Politics and technology have been intimate bedfellows for quite some time. One very well-known example is the Manhattan Project, which led to the creation of the atomic bomb during World War II. The military potential of nuclear fission was first "pushed up" onto the political radar because nuclear technology seemed to answer the primary need of the Allied commanders – to bring the war to an end as quickly as possible. There was a clear affinity between the possibilities offered by the technology and the political needs of the day. I will refer to such cases as illustrating a "bottom-up" approach, because pre-existing technology is appropriated to serve a political end that it wasn't intended for, a purpose that was likely not even anticipated at the time the technology was first designed. In the next chapter, I'll describe the more rational "top-down" path – situations where technological systems are deliberately designed to serve a particular pre-defined political purpose.

I'd like to take four remarkable events – or developments – to illustrate the bottom-up path. The terrorist act of September 11th, 2001, the IBM/Holocaust case study, and the use of video cameras and the Internet to advance human-rights activism are all examples that achieve a tight Human-tech fit between policy aims and technology.

These examples are striking because they show that Human-tech thinking at the political level can make use of existing technologies from below to serve remarkably diverse political imperatives. An airplane can be a safe and efficient mode of transportation, or it can be a weapon of mass destruction; punch-card technology can be used to conduct a census or to orchestrate a campaign of genocide; video cameras can be used to document human-rights abuses or to create child pornography; the Internet can be used to convince the world that human rights are being violated or that the Holocaust never really happened. If you can create a tight Human-tech fit between a technological system and human nature, then you have a much better chance of achieving your political objectives – for good or ill.

THE ART OF WAR: 9/11

Military strategists – at least the clever ones – have known about the power of this bottom-up path to Human-tech thinking for centuries, even millennia. The leader of the pack in this respect is undoubtedly Sun Tzu, who wrote *The Art of War* sometime between 500 and 300 B.C.E.[1] In his classic work, Sun Tzu wrote that "a wise general strives to feed off the enemy" and "good warriors seek effectiveness in battle from the force of momentum."[2] Just as an expert judo wrestler uses the weight and strength of his opponent to his own advantage, Sun Tzu advocated that generals and warriors manipulate the resources and capabilities of their enemies to create a momentum that will advance their own purposes.

It's an elegant idea, and it became a landmark in intellectual thought because it captures something both deep and enduring about human nature at the political level, something that transcends the technology of any particular time and the idiosyncratic details of specific wars or battles. Why waste your own resources when you can use those of your enemy? It's a simple but profound insight. Not only does it have the advantage of efficiency, but it also has the benefit of generalizability: the strategy can be used by all kinds of political groups in all kinds of situations – including terrorists intent on bringing the most powerful nation in the world to its knees.

Despite the fact that *The Art of War* was written more than two thousand years ago, an argument could be made that the September 11th terrorists planned their diabolical but effective attack by taking a page out of Sun Tzu's playbook. They didn't have the resources to design and create a system of their own to achieve their political aims, but they were ingenious enough to identify an existing complex technological system that was (unintentionally) exactly tailored to their destructive purpose. Every single detail seemed to provide a seamless design fit. At the organizational level, the U.S. Immigration system was "designed" to be lax enough to let the terrorists into the country easily using student visas, despite the fact that several of them were known enemies of the United States. At the psychological level, the commercial jet airplanes were designed so that the psychomotor skills required to fly them could easily be acquired with some basic flying lessons. At the physical level, the random baggage searches of airport security systems were designed in such a way that it remained possible to smuggle deadly objects on board. The weapons themselves – the innocuous-appearing box cutters – were designed to cut cardboard but could be adapted for efficient killing. The long transcontinental flights from the East Coast to the West meant that the planes were designed to carry large loads of fuel, thus guaranteeing the terrorists a free, potent source of explosive. The largely empty flights made it a simple matter for a handful of terrorists to take control of the crew, the passengers and the airplane. The design of the tall, imposing World Trade Center towers made them tragically easy to recognize and find, and therefore easy to fly into: perfect targets. The materials that were used to construct the twin towers were designed to buckle and yield to a raging inferno fed by the impact of a fully fuelled commercial airplane. The enormous size and weight of the mighty buildings were designed to assure mass destruction upon their collapse. And such a collapse ensured that a vast number of the people who worked in the two towers every day would become victims of mass murder, no matter how valiant the efforts to evacuate the buildings and save lives. Finally, at the political level, the timing of the event – early in the morning of a bustling New York business day – was "designed" so that the mass murder and destruction would inevitably be noticed and recorded for posterity by the media, with all the attendant consequences, while the location of the towers – the fact that they were in Manhattan, the financial epicentre of Western capitalism, and filled with financial and trade workers from around the world – made them an impeccable ideological target. Even the cleverest of system designers would have been hard pressed to create a technological system that could have resulted in a better multi-level fit with the terrorists' political aims. Knowingly or not, the terrorists' plan embodied Sun Tzu's ageless lessons, illustrating the bottom-up path of using your enemy's technology in the service of your own political aims.

The end result shows the incredible "amplification effect" that can be achieved by co-opting for destructive purposes a technological system that was designed to leverage human nature and the laws of physics for purely benign uses: 10 million square feet of office space collapsed in about fifteen seconds, killing thousands of people, inflicting $30 billion of damage in direct costs, and a profound change in geo-political thinking.[3] Yet, horrific as it was, September 11th wasn't the first or even the most egregious example of a brutal regime tailoring existing technologies at the political level to create widespread human suffering and destruction.

GENOCIDE BY AUTOMATION: IBM AND THE HOLOCAUST

Long before the days of personal computers, information was processed by means of paper punch cards and mechanical sorting machines of the type that are still employed in some elections today. They were originally invented by Herman Hollerith in 1884 to calculate demographic statistics.[4] Although primitive by the standards of today's computers, Hollerith machines were a tremendous breakthrough in their day, representing a huge improvement over manual methods of compiling, sorting and cross-indexing census information. In fact, the technology was so efficient that it helped the U.S. Bureau of Census save $5 million – one-third of its budget – during the 1890 national census. Not only were Holleriths cheaper, but they were powerful as well. Whereas the 1870 census merely asked five questions of citizens, the 1890 census asked 235 questions – a 47-fold increase. More information for less money – a guaranteed recipe for business success.

Not surprisingly, the use of punch-card technology quickly spread to many other applications, particularly those requiring actuarial and financial calculations. The new Hollerith technology was a financial hit – so much so that it was acquired by a company that eventually became known as International Business Machines, arguably the most powerful multinational corporation of its time. IBM enjoyed tremendous profits by deploying Hollerith machines and cards in virtually all its subsidiaries overseas. Money-making applications were discovered in diverse industrial sectors in many countries – including Nazi Germany.[5]

It appears that Hitler made efficient use of their potential. Hollerith machines and cards helped the Nazis search through vast public census and genealogical records to identify quickly the Jews who were to be exterminated. To do the same job manually without the lightning-fast automation would have taken many more people and a much longer time. The genocide couldn't have proceeded as swiftly and as comprehensively as it did without the IBM-owned technology.

But the utility of Hollerith machines didn't end there. The same technology was also used to catalogue the prisoners at Auschwitz and the other infamous Nazi concentration camps. The custom-designed punch cards consisted of many columns, each of which had several numbers that represented prisoner information. Information was coded on the cards by punching out a number in each column of the card, creating a hole. Each column represented a different prisoner attribute, and the presence of a particular hole in that column had a specific meaning. For example, columns 3 and 4 identified each prisoner as belonging to one of sixteen categories: hole 3 denoted a homosexual; hole 8 denoted a Jew; hole 12 denoted a Gypsy. Column 34 was used to code the prisoner's "Reason for Departure": hole 2 denoted a transfer to another concentration camp for continued hard labour; hole 3 denoted death by suicide; the ominous hole 6 denoted "special handling" – the Nazi euphemism for extermination by gas chamber, shooting or hanging. Even the five-digit number that concentration camp prisoners had tattooed on their forearm – an enduring icon of the Nazis' genocidal goals – had a technological connection. It corresponded to a number on a Hollerith punch card that held information about that particular prisoner. Once all this information was coded, the Nazis were able to coordinate the tremendous number of activities that it took to consummate their "Final Solution" with the speed and accuracy that only automated technology was capable of providing.

The Hollerith machines and cards were also used to inventory Jewish assets, to keep up-to-date statistics on the number of Jews killed, and to satisfy labour requirements by matching the professional and language skills of the prisoners with the needs of the various concentration camps. The same tireless and obedient technology was also used to transport prisoners, schedule trains and facilitate all the other logistical planning across dozens of cities in over twenty countries and territories.

The exploitation of the IBM technology provides another dark example of our ability to adapt technology to answer many purposes including virtually any political agenda. "Good" designs can be used for "evil" purposes; the Hollerith machines and cards were a tremendous technological innovation that helped people solve pressing problems in a far more effective way than had been previously possible, but the cruel irony is that they were just as effective for murdering six million Jews as they were for conducting a national census.

TECHNOLOGY FOR HUMAN RIGHTS:
LOW-TECH CAN BE HUMAN-TECH

Technology itself is value-neutral, neither good nor bad – there is no soul in the machine. And Human-tech thinking – which recognizes the fit with human nature at a political level – can be used for oppressive purposes; but it can equally be used to advance humane values. It's interesting to see how technologies that were developed for entirely different purposes have been imaginatively co-opted to promote human rights around the world. It seems that human nature is particularly resilient at the political level in countries where human rights are routinely violated.

Most people who live in countries with democratic forms of government tend to think of the "political" as a relatively specific and limited category – it involves activities such as advocacy, voting, open debate, educating and so on. However, those who have spent time in countries ruled by oppressive dictatorships will have seen, at first hand, a profoundly different side of human nature at the political level, unlike anything normally on view in Western democracies. As the saying goes, all politics is local, and being a political activist in countries where human rights are routinely violated puts an entirely different set of constraints on what a Human-tech technological system – in which technology is exquisitely adapted to human needs and capabilities – needs to look like.

I learned something about these constraints – at second hand – because I have an uncle who was a political dissident in Portugal during the fascist dictatorship of Salazar.[6] The challenges he faced would be instantly recognizable to dissidents struggling against dictatorships in all parts of the world. Before the 1974 revolution that ushered in democracy, if you were found to engage in any political act deemed threatening to the state, then the typical punishment was arrest, interrogation, torture and imprisonment, possibly for years at a time, a fate suffered by my uncle and countless others. So secrecy was an essential over-riding concern for human-rights activity. There were also specific constraints on particular political activities. For example, in order to disseminate information or engage in networking and debate, you had to keep in mind that phone lines were routinely tapped; you didn't dare discuss any sensitive information over the telephone. Activists resorted to communicating with each other in extremely roundabout ways to escape surveillance. One popular method, with a historically solid genealogy, was to pass messages by putting ads in newspapers, using a numerical code to identify the key words in the message. For example, the code 7234 might point someone to look at page 7, column 2, line 3, word 4 of a newspaper. Politically sensitive messages could be painstakingly decoded, one word at a time, using such methods, but would be unnoticed by the casual newspaper reader and unintelligible to censors of the dictatorial government.

Communicating directly rather than via means that were censored or monitored was essential, both in and out of jail. The means that jailed dissidents used to communicate with each other while in prison shows just how far our everyday technologies can be imaginatively twisted to serve political purposes. Prisoners were forbidden to speak to each other, monitored by guards, and physically segregated in different parts of the jail, and yet they still managed to communicate with each other on a regular basis by assembling a primitive information system out of the few objects and activities that they had access to while in jail. The process began by writing a message on cigarette rolling papers in a script so small it was almost unreadable by the naked eye. The cigarette papers were put into a makeshift package that the prisoners created by co-opting a plastic bag and sealing the plastic with a flame from a cigarette lighter. The tiny message in a bag was concealed in the prisoners' dirty laundry and sent to be washed. Because the paper was tightly sealed in plastic, water couldn't get into the bag and the message usually survived the washing intact. Afterwards, the clean laundry was hung outdoors to dry on a clothesline that would, at some point, be accessible to the other prisoners who could then "read the laundry" – reach into the drying clothes, pull out the little plastic bags, take them to their cell, and there read the enclosed message, all without the guards knowing what was going on.

Outside the prisons, networking was also a challenge because any public meeting involving more than three people was legally defined as a subversive political gathering, providing an excuse for the state police to round up people and dish out the usual punishment. Dissidents trying to escape detection wore disguises, used passwords, and met up with fellow dissidents whom they had never seen before using covert identifying signals. For example, you might be told to go to the town square at noon, look for the tall guy wearing blue overalls who was eating an orange, and utter a particular phrase to establish contact. And ways of organizing that transcended distance were also critical. Networking with dissidents beyond your local circle was a challenge, because suspected dissidents were routinely followed by the state police. Bicycles became a very popular tool for transportation; they were cheap, they didn't stand out, and perhaps most important of all, they made it easier to take circuitous, back-road routes to reduce the risk of being tailed and caught.

Good communication – the give-and-take of information – is fundamental to any democratic political body, so of course it was also important to the dissidents' political objectives. But since there was no such thing as free speech, and all the media were state-controlled and subject to censorship, dissidents resorted to cheap, portable, clandestine ways of influencing public opinion, publicizing events and providing information, such as how many young soldiers were being maimed or killed in the futile colonial wars that the Portuguese dictatorship was waging overseas in Mozambique and Angola. Ingenious makeshift printing presses were created, small enough to be easily hidden and quickly moved if you suspected the state police were closing in. My uncle had a homemade press that was jury-rigged from domestic materials: a small wooden frame, fabric from a nylon shirt, and a small roller. Using ink or wax, he would make copies from a master message – one sheet at a time – and then the newsletters would be distributed on the sly to interested (and trustworthy) readers.

The technologies I've been describing – numerical codes, messages on cigarette papers in tiny plastic bags, secret signals, bicycles and stealth printing presses – were co-opted because they had an affinity with the activists' needs in an oppressive political environment. As such, they're examples of Human-tech thinking at the political level, tools that played an indispensable role in organizing a political countermovement that eventually overthrew Portugal's Fascist dictatorship in 1974. They are also marvellous examples of how low-tech Human-tech can be. But that doesn't mean there isn't room for improvement – advocating human rights using such primitive means is clearly a very slow, uphill battle. It was no accident that Salazar was prime minister of Portugal for thirty-six consecutive years.

THE EYES OF THE WORLD ARE WATCHING NOW:
VIDEO CAMERAS AND THE INTERNET

Today, human-rights activists are using very different and far more powerful tools to advance their cause, and two newer technologies in particular have had a transformative effect on political activism, playing an extraordinarily effective role in the spreading of ideas and democratic discourse, particularly amongst the global youth movements. Co-opting video technology and the Internet for political communication is a brilliant example of Human-tech thinking because there is a perfect fit between the design of those systems and the urgent political needs of many in oppressed classes or countries.

Witness is a non-governmental organization co-founded by the rock musician Peter Gabriel, which provides video cameras and field training to human-rights activists.[7] As the Rodney King beating in Los Angeles and the World Trade Organization protests in Seattle showed, video images of human-rights abuses are a uniquely powerful way of attracting the eyes of the world via repeated TV, grassroots advocacy, and Internet broadcasting. Witness has organized more than 150 partner groups from fifty countries, and the videos obtained have been used in legal proceedings, for education, in news broadcasts, in web broadcasts and to counterbalance reports that governments make to the United Nations about their human-rights records.

Cameras can be used to record events directly, bypassing government censors. They're also relatively anonymous – the camera's operator is invisible – satisfying the criterion of covertness. The video medium is also fast because it can be transmitted very quickly, and the images can be broadcast all around the world, swiftly maximizing the number of people that can be reached. And finally, the vivid nature of the images provides a tangible, memorable, durable and emotionally compelling record of collective passion and protest or atrocities being committed, thereby increasing the potential to influence public opinion. It's far more difficult to get away with tyranny and torture when you know the world is watching your every move.

Not that there's anything inherently virtuous about video cameras; the same technology can be used to create child pornography for pedophiles. But in this particular political context, possibilities offered by video cameras have a profound affinity with the activists' need to communicate instantly and widely, far more than did the makeshift printing press concocted by my uncle. Just as the Fender Strat provides a snug physical fit with the size and shape of a guitarist's body, video technology is beautifully tailored to the political form of activist organizations.

Human-rights organizations have been quick to make use of Human-tech technology. My uncle needed to communicate sensitive information directly to fellow dissidents, and to network covertly with geographically distant supporters; human-rights activists today face the same political imperatives. But instead of relying on bicycles, numerical codes, disguises or passwords, organizations in every corner of the world are turning to the Internet to promote their cause.[8] Because it's a highly decentralized technology, activists can communicate with each other directly, thereby minimizing the likelihood of censorship and control by authoritarian government authorities. And because the Internet doesn't know or care about national and political borders, it effectively overcomes the "accident of geography." Activists network and coordinate across huge distances. Thanks to the miracle of digital technology, information is transmitted very quickly – an essential advantage to activist groups, particularly in times of crisis. The relatively low cost of Internet technology makes it available as a tool to a broad user group, although access is still a problem in some parts of the world, despite Internet cafés now scattered widely through distant deserts and mountains and jungles. And, although electronic surveillance by intelligence agencies is on the rise, anonymity is largely safeguarded since no one need reveal their true identity to use the technology. Finally, the Internet is also very efficient: the amount of work it takes to send one message is about the same as that required to send many messages. That sure beats taking a long meandering bicycle ride through back roads to attend a covert political meeting with a mere handful of activists.

Obviously, the Internet is available to all comers, and when co-opted as a political tool can be equally effective in helping white supremacists and other hate groups spread venomous messages to a growing audience (after all, evil is in the eye of the beholder). My sixty-year-old uncle – who has learned to surf the web, and has an e-mail address of his own – is delighted to see that activists now have much more sophisticated technologies at their disposal than he once did, technologies that fit their needs as seamlessly as a custom-tailored shirt or the Fender Strat;

and that Human-tech thinking at the political level can indeed be exploited to advance human rights.

Human-tech thinking at the political level can co-opt existing technologies from below to serve political imperatives. The September 11th terrorist acts, the IBM/Holocaust case study, and the use of video cameras and the Internet to advance human-rights activism are all examples of this bottom-up path to achieving a tight Human-tech fit between policy aims and technology. But if that were the whole story of Human-tech thinking at this level, we would need only to sit around with a watchful eye on the latest technologies to see which ones can be leveraged for political purposes. As I'll show in the next chapter, Human-tech thinking at the political level also involves taking a deliberate and systematic top-down approach, making decisions to create a design that can govern and manage complex technological systems safely and effectively.

9

Political Imperatives II:
Safeguarding the Public Interest

POLITICS AS DESIGN: POLICY AIMS,
LEGAL REGULATIONS, AND BUDGET ALLOCATIONS

It may seem odd to think that a *political* decision could be a deliberate part of "system design." Most of us wouldn't see any similarity between the design decisions that led to the creation of the Reach toothbrush and the policy decisions made by politicians, public servants and non-governmental organizations that led to the creation of the Kyoto protocol. You can hold a toothbrush in your hand and use it directly to complete the task for which it was designed. Policy decisions can't be held in your hand. For instance, an international accord is an abstract cultural constraint, and although it can be physically written down on paper, you can't pick up and use that piece of paper to achieve the goals that the legislation was intended to fulfill – at least, not in any direct sense.

The differences between the concrete physical world and the abstract political world are obvious, of course, but behind those real differences are subtle similarities that are crucial from our Human-tech perspective. The design decisions that went into the Reach toothbrush and the policy decisions that created the

Kyoto protocol are both system design "levers," in the sense that they are tools that shape human behaviour in ways that affect the success of a system, albeit not on the same scale. Think about a specific contrast: just as the shape of a toothbrush can affect how easy it is for you to brush your back teeth, the form of product-liability legislation can affect how easy it is for a company to change the design of their product after an accident, without having to worry that such a change might be used against them in a lawsuit.[1] Moreover, for either design to be useful or effective, it's important that it be tailored to what we know about human nature. Just as toothbrushes should be based on an understanding of the size and shape of our hands, mouths and teeth, legislation has to be predicated on an understanding of the socio-political forces that motivate people in the area in question. And in both cases, the overarching objective is to satisfy a human or societal need – whether to achieve effective dental hygiene or provide safer health care. These commonalities between the physical and the political emphasize that system design isn't only about the construction of "stuff." Complex technological systems span both hard and soft technology, and can include the construction of less tangible entities, such as government legislation.

The conceptual, slippery, non-physical nature of system design at the political level isn't the only challenge to implementing a Human-tech approach on this grand scale. Predictability is the other problem. Designing a system *to* something requires a minimal amount of certainty. We have to have some knowledge about the "something," or we don't have a clue how to tailor the design of the system to achieve a good fit. From this vantage point, the very idea of designing complex technological systems to have a top-down affinity with human nature at the political level may seem at best ambitious, and at worst a complete waste of time. Politics seems unpredictable. How can you tailor the design of a system to conform to the characteristics of a phenomenon that seems chaotic?

Admittedly, there's a lot that's unknown about human political behaviour. We don't have the functional equivalent of an anthropometry handbook where we can look up a bunch of numbers that describe human nature with objective precision; the range of political behaviour is far more complicated and dynamic. Thus our understanding of the structure of political behaviours is far more elusive than our understanding of how the body works.[2] As a result, tailoring the design of a system to human nature is an immense challenge at the political level; it's much more difficult to figure out if a new public policy initiative will be well received by voters than it is to determine if the controls on a new lathe will be easy for people to reach. But researchers in political science, sociology and law have amassed a surprising amount of knowledge about people as political creatures.

Henry Kissinger once said that if you wanted to know how politics worked on the eve of the twenty-first century, all you had to do was read Machiavelli's 1532 classic, *The Prince*.[3] According to Kissinger, it's all there; nothing much has really changed in the last 470 or so years. Some might say that Kissinger's statement reveals more about his personal political tactics than it does about human nature in general,[4] but it's undeniable that we have, over the centuries, gathered a great deal of timeless, pragmatically useful knowledge about human nature and politics.

Although most of us don't think of political decisions as "system design" interventions, they truly are, because choices about policy aims, budget allocations and legislation have an impact on how likely it is that a snug Human-tech fit will be achieved, not just at the political level itself, but also at the organizational, team, psychological and physical levels of human nature. Indeed, the Human-tech idea of satisfying human and societal needs at the political level by seeking a closer affinity with human nature has been around for a long time. In 1759, seventeen years before he penned *The Wealth of Nations*, Adam Smith

– the seminal systems-approach guru, who is usually thought to
have put economic self-interest above all else as the strongest
force in society – authored the lesser known but equally brilliant
Theory of Moral Sentiments. In it, he wrote:

> The man whose public spirit is prompted altogether by
> humanity and benevolence, will respect the established powers
> and privileges even of individuals, and still more those of the
> great orders and societies, into which the state is divided. . . .
> When he cannot establish the right, he will not disdain to ame-
> liorate the wrong; but . . . when he cannot establish the best
> system of laws, he will endeavour to establish the best that the
> people can bear.[5]

In my terms, this is *top-down* systems design, which puts great
emphasis on the relationships between system elements. And to
turn Adam Smith's idea into action in modern times, we need
to identify the system design levers at the political level that are
relevant to the success of complex technological systems. Three
levers identified by political scientists include policy aims, legal
regulations and budget allocations.[6] *Policy aims* represent value
judgments and are perhaps the most important of all because
they provide an explicit direction – a compass heading – to ori-
ent the construction and operation of the system as a whole,
providing a criterion for aligning all of its elements. Because
technology is changing and increasing in complexity at such a
frenetic pace, we don't usually have time to reflect on, let alone
question, the rationale behind technological innovations that
affect all of our lives. We just ride the wave of technological
"progress," hanging on for dear life. As a result, we frequently
forget that the way technology is used in society really is under
human control. *Somebody* has to make decisions, even if only
implicitly, about which technologies should be introduced and
for what ends. After all, technology itself only offers possibilities;

it has no intrinsic moral value. As John Ralston Saul observed shortly after September 11, 2001, "Machinery. . . . cannot lead. It is mere technology and it can be used by different humans for whatever purposes suit them. . . . A passenger jet may indifferently deliver you to a beach holiday or become a deadly missile. An exacto knife may cut a piece of paper or a throat."[7] Policy aims are the mother of all system design decisions.

Legal regulations can also have an important top-down influence on a complex technological system by creating the penalties and rewards that shape the behaviour of organizations, teams and individuals, as we saw in the aviation and health-care sectors. In the case of health care, the "invisible hand" of legal liability can be destructive, leading to death by medical error.[8] The fact that legislation was passed to ensure that the 500,000 reports in the ASRS database aren't open to legal discovery and therefore can't be used as fodder for lawsuits is a benevolent invisible hand, which in closing that door has contributed to a tremendous improvement in aviation safety. Countless lives have been saved, showing that laws can have positive effects if they're deliberately and knowingly designed with human nature in mind. If policy aims are a compass, then legal regulations are the engine that powers a complex technological system in the desired direction.

Finally, *budget allocations* are also critical because they provide fuel for the engine. A government can pass legislation requiring that every hospital have an incident-reporting system to improve patient safety, but unless each hospital is allocated enough money to implement and maintain such a system, the new law will likely be ineffectual and the policy aim will remain unfulfilled. Determining how taxpayers' scarce money should be spent is a third important "system design" lever available to governments.

These three political factors – policy aims, legal regulations and budget allocations – may seem like bureaucratic minutiae of concern only to government bean-counters, not to the average citizen. Nothing could be farther from the truth, and as citizens we should sit up and take notice, given the increasingly complex technological world we all live in. Top-down political design interventions are literally life-and-death decisions – modern-day invisible hands that can touch and change the quality of *all* our lives. Sadly, this isn't just hyperbole.

DRINKING WATER OR DEADLY POISON?
THE WALKERTON E. COLI OUTBREAK

Walkerton, Ontario, used to be an ordinary and relatively anonymous small town located northwest of Toronto, typical of most rural towns the length and breadth of North America. If you remember Mayberry – the sleepy, neighbourly town that served as the setting for *The Andy Griffith Show* on TV – then you have a picture of what Walkerton was like: nothing too exciting ever happened, everyone knew everybody, and there was a strong sense of community. But in May 2000 the town was blindsided by a disaster that shocked and terrified all of its inhabitants. The name Walkerton became forever linked in Canadian history with death by poison after the public water supply system became contaminated with lethal E. coli bacteria.[9] In a town with a population of 4,800, seven people died and an estimated 2,300 became sick. Some people, especially children, are expected to experience lasting damage to their health. The total economic cost of the tragedy was estimated to be over $64.5 million. If a comparable calamity had struck in a city the size of Paris or New York, the number of dead would have been about 14,000, and 4 million would have become ill.[10]

In the aftermath of the deaths and illnesses, citizens were terrified of using tap water to satisfy basic human needs, such as bathing, cooking and drinking. Many still are. Those who were infected or lost loved ones suffered tremendous psychological trauma; their neighbours, friends and families were terrorized by anxiety.

But what made Walkerton swiftly notorious was the fear that spread like a brushfire across the country. People didn't understand how a town's drinking water could become infected with killer bacteria in this day and age of technological progress, and worried that the same thing could happen in their town or city. News of the events of May 2000 also rippled out across the world as the unassuming Walkerton residents became unwilling participants in an international media circus that descended on the town. Headline news stories continued unabated for months in newspapers, radio and television. Even CNN sent a crew to sleepy Walkerton to cover the events. A fury of political pressure was brought to bear on the Ontario provincial government – a previously popular Conservative Party government that had won two consecutive elections by a landslide majority vote while running on what the party called a "Common Sense Revolution" platform of smaller government and fiscal responsibility – as the public fought to identify who was responsible and what was being done to safeguard the drinking water systems all over the province.

Eventually, the provincial government caved in to the public outcry and the intense media scrutiny by appointing an independent commission to conduct a public inquiry into the causes of the disaster and to make recommendations for change. The commission was given wide powers of investigation, and collected as many as one million documents from the provincial government alone, in addition to thousands of documents from other sources. In the end, about 200,000 government documents were scanned into an electronic database. Ninety-five days of publicly televised hearings were held over a period of nine months, during which 21,686 pages of transcripts were generated and 447 exhibits, containing over 3,000 documents, were introduced as evidence.

A total of 114 witnesses were examined, culminating in the politically devastating interrogation of Mike Harris, then premier of Ontario.

Before I describe the dramatic sequence of events that led to the tragedy, I need to provide some background information about the key players. The town's water system was operated by the Walkerton Public Utilities Commission (WPUC), under the supervision of its general manager, Stan Koebel – a plain family man respected and trusted by his neighbours, but not highly educated, who had started working at the WPUC straight out of high school as a teenager. Stan's brother, Frank – also a simple small-town guy without any university education, who had lived his entire life in Walkerton – was the foreman at WPUC. The Ontario Ministry of the Environment is the government body with oversight responsibility for the operation of water systems throughout the province. Other players in the drama would be the Ontario Ministry of Health and the Bruce Grey Owen Sound Health Unit – the regulatory body for those counties headed by the local medical officer of health. Finally, there is A&L Canada Laboratories, a private company that had been contracted by WPUC to analyze their water samples.

So what exactly happened in the small town of Walkerton in May 2000? The first contributing factor seemed innocuous enough at the time: between May 8 and 12, the town experienced day after day of unusually heavy rainfall. The relevance of that meteorological rarity becomes clearer when we learn that Well 5, one of three wells feeding the Walkerton water system, was located near a farm that had recently been spread with cattle manure which, naturally, contains E. coli and campylobacter bacteria. The potentially lethal contaminants were carried by the heavy downpour into Well 5 on or about May 12.

The WPUC staff were supposed to take daily measurements of chlorine levels in Well 5, between May 13 and 15, but failed to do so. On May 15 water samples were collected, although not from the required locations. (I will describe the reasons for these procedural violations later.) These samples were sent to A&L Labs for testing in the normal way. Two days later, on May 17, A&L Labs advised Stan Koebel in writing that the test results of those samples showed massive contamination of the water due to potentially lethal E. coli and campylobacter bacteria. Stan later said that he was very busy when the test results came back and didn't look at them closely until a few days later. In the meantime, information pointing to a deadly public health threat lay unread on his desk.

The next day, May 18, the first signs of trouble appeared in the community. An unusual number of people started becoming sick. To appreciate what the citizens of Walkerton were already going through at this point, just consider what E. coli does to the human body.[11] It usually takes about three to four days for people to become sick after ingesting the bacteria. The initial symptom is raging diarrhea, which can last as long as ten days. Within twenty-four hours the diarrhea becomes bloody. At the same time, people can experience excruciating stomach pains as the tiny bacteria begin to wreak havoc inside their bodies. Local health-care providers were surprised when they learned that several people were experiencing these unusual symptoms, so the WPUC was contacted to see if the water might be the source of the problem. They were told that the water was OK.

On May 19, the scope of the outbreak grew, more people, particularly children and the elderly, became sick. A pediatrician contacted the Bruce Grey Owen Sound Health Unit because she suspected that E. coli might be the culprit. The health unit began its investigation on the same day. Twice, they phoned Stan Koebel to ask about the state of the water supply. Both times, they were reassured that the water was safe. In addition, Stan failed to disclose the distressing test results that he had received from A&L Labs. It turned out later that he had been operating the third well – Well 7 – without the chlorinator running, and thinking it was this violation of policy that had caused the contamination, he was trying to cover up the problem in the hope that life would return to normal before his violation was detected. Things didn't quite turn out that way.

The repeated questions about the quality of the water worried Stan Koebel. That same day, May 19, he began to flush and super-chlorinate the Walkerton water system to destroy any possible contaminants. Gradually, the chlorine values began to climb into the safe region, which gave him confidence that the quality of the water wouldn't pose a threat to public health. He would soon learn that it was already too late to avert a deadly disaster.

On May 20, the first preliminary test for E. coli infection came back positive – a child with bloody diarrhea – suggesting a local source of potentially lethal contamination. Once more, the possibility of a problem with the water supply was considered, and the health unit went back to Stan Koebel twice more. He reported – truthfully this time – that the chlorine values were now at an acceptable level, but once again he failed to disclose the adverse test results from the May 15 samples. Taking him at his word, the health unit informed the public that the water supply wasn't the source of the burgeoning epidemic.

Now, however, a crack appeared in the wall of lies hastily erected by Stan Koebel. An anonymous WPUC employee placed a call to the Ontario Ministry of the Environment to report the test results from the contaminated sample that had been collected on May 15, almost a week earlier. But when the ministry contacted Koebel, they were told that the only negative water sample was collected from a construction site – another lie. Because the ministry had never been sent the test results from A&L Labs, they had no way of independently confirming whether Koebel was giving them an accurate or complete account of the quality of the drinking water. That afternoon, the health unit, in mounting concern, contacted the local medical officer of health, who proceeded to take over the investigation.

On the next day, May 21, the health officials' worst fears were confirmed – the preliminary positive test for E. coli from May 20 was officially confirmed and a second preliminary positive report was received, proving that the town of Walkerton was indeed facing a potentially deadly outbreak. A "boil water" advisory was issued by the health unit – although that advisory didn't reach everybody concerned because it wasn't as widely publicized as it could or should have been. The ministry and the local health unit both contacted Stan Koebel, who replied with a by then familiar refrain – the water was fine – still without disclosing the negative sample results from May 15. This time, however, the health unit, suspecting that the water quality might indeed be the cause, went ahead and collected its own samples from the Walkerton water distribution system.

One full week after the negative water sample was first collected, the outbreak reached an emotional turning point. The ministry began its own investigation, and Stan Koebel finally turned over all of the documents he had, including the test results from the May 15 samples showing massive contamination of the public water supply from lethal E. coli bacteria. On May 22, the first person died from the outbreak – sixty-six-year-old Lenore Al, a retired library worker who left a husband, three sons and ten grandchildren. Eventually, six more deaths would follow, and 2,300 people would become ill.

Clearly, human actions played an important role in the failure of the Walkerton public drinking-water system – an example of a complex technological system. Stan Koebel's miserable cover-up and deceptions influenced the course of events, subverting the technology's ability to deliver safe drinking water to the residents of Walkerton. But as the subsequent investigation revealed, Koebel was not solely at fault. The system in which he worked had glaring flaws, which contributed significantly to the disaster. Indeed, during the subsequent inquiry hearings, when lawyers representing the provincial government tried to pin all the blame on Stan Koebel, their contention was rejected out-right by the widely respected inquiry commissioner, Dennis O'Connor: "It is simply wrong to say, as the government argued at the inquiry, that Stan Koebel or the Walkerton PUC were solely responsible for the outbreak or that they were the only ones who could have prevented it."[12] To understand why Commissioner O'Connor didn't buy the provincial government's heavy-handed legal tactics, we have to understand the other contributing factors that led to the events of May 2000.

The heavy rains and the use of cattle manure on local farm fields combined to introduce the contaminants into Well 5. But the unusual weather wouldn't have had any negative impact without four other pre-existing factors.

Two of these four contributors were *environmental factors*. It turned out that Well 5 was a shallow well, drilled to a depth of just 15 metres. Walkerton's two other wells – Wells 6 and 7 – were drilled to depths of 72 metres and 76 metres respectively. In addition to its shallowness, Well 5 was drilled in an area where the bedrock was highly fractured and porous. The soils, sands, silts and clays covering the bedrock were also quite shallow in that area. As a result, the natural filtration processes that typically cleanse rain-water as it drains on its way to a well could never have been very effective – in short, Well 5 had been particularly vulnerable to con-tamination from surface water by runoff from rainfall, right from the very day it was first constructed in 1979 – a full twenty-one years before the outbreak.

The other two factors that contributed to the vulnerability of Well 5 to surface contaminants were *equipment related*. Wells known to be exposed to the risk of contamination by surface water, like Well 5, are normally required to have continuous chlorine monitors installed – such a monitor would automatically shut down the pump at Well 5 when the contamination entered the system. But for reasons that I'll discuss later, Well 5 didn't have a continuous chlorine monitor. The second equipment-related factor was the absence of a working chlorinator in Well 7 between May 3 and May 19. With Stan Koebel's full knowledge, Well 7 had been supplying unchlorinated water to the distribution system for several days during this period, in clear violation of provincial requirements. This action didn't contribute to the contamination, but it did cause Koebel to deliberately conceal the negative test results in an attempt to cover up his violation.

But the environmental and equipment-related factors weren't the only complications. The poor technical and operational management of the Walkerton Public Utilities Commission, under the supervision of Stan Koebel, also played a crucial role in shaping the events of May 2000. The failure of the WPUC to take chlorine residual measurements at Well 5 between May 13 and 15 and the active concealment of the results of the water samples collected on May 15 were critical. The astonishing key point, however, is that these were not unusual occurrences: "*for more than 20 years,* it had been the practice of [WPUC] employees not to measure the chlorine residuals on most days and to make fictitious entries for residuals in the daily operating sheets."[13] And the improper operating practices at the WPUC didn't end there. Workers had also been routinely and deliberately misstating the locations from which the water samples had been taken, collecting insufficient numbers of water samples, failing to use adequate doses of chlorine, and even submitting false annual reports to the Ministry of the Environment.

What contributed to this long-standing pattern of negligence? First, the WPUC management, particularly Stan and Frank Koebel, had never been adequately instructed about the technology they were in charge of. Both Koebels had received their water operator's certificates through a grandfathering process. Neither of them had ever been required to complete any training or pass any examinations whatsoever before receiving their certification. They each believed that the sources for the Walkerton water system were generally safe. Indeed, both men routinely drank untreated water at the well sites because it tasted better. Even over the critical weekend of May 20–21, after he knew of the adverse test results, Stan Koebel "continued to drink water from a fire hydrant and a garden hose, and on May 22, he filled his daughter's swimming pool with municipal water."[14]

Although this doesn't justify his actions, it's also interesting to note that the Walkerton residents had sometimes told Stan – their neighbour and friend – that the water tasted too much like chlorine, so in an effort to respond to the wishes of his small, tight-knit community, he had reduced the amount of chlorine he introduced into the system.

These facts put the events of May 2000 in a somewhat different light. Not knowing them, we might be tempted to suspect that Stan Koebel was a deliberate saboteur of the water system. But no sane saboteur would put his own life or that of his daughter at risk. And a review of the technical and operational management of this complex technological system shows that neither of the Koebel brothers had a full appreciation of the health risks to which they were exposing themselves, their friends, their neighbours and their own families when they distributed under-chlorinated water, nor of the specific risks of bacterial contaminants. In fact, both admitted to not even knowing what E. coli was, let alone that its presence in drinking water could be lethal, and Stan Koebel in particular – attempting to do his job well – had tried to be responsive to his neighbours' requests for water with less of a chlorine taste.

The fact that violations of procedure and falsification of records were normal at the WPUC also contributed to the situation. Everyone did it and thought nothing of it. Samples were often collected at convenient sites, including the operators' homes or the WPUC workshop, and then simply labelled according to where they were supposed to have been taken. When asked why he and his brother followed and allowed such practices, Stan Koebel replied: "Simply convenience."[15] And even though several Ministry of the Environment inspections conducted in the 1990s had detected and pointed out these deficient practices, and Stan Koebel had made promises to rectify the problems, he never did so.

These additional pieces of the puzzle reinforce a point I made before – management truly matters in maintaining the safety of a complex technological system. The attitudes and actions of the Koebel brothers "show a serious disregard for [Ministry] requirements and repeated failures by Stan Koebel to do what he said he would."[16] These actions are inexcusable. But Commissioner Dennis O'Connor – a thoughtful judge with great integrity – didn't accept them as the final explanation for the Walkerton outbreak. After all, several questions remained to be addressed: How did the WPUC get away with this kind of behaviour for so long? Why weren't these deviations detected and corrected *before* lives were lost? Did the government – local and provincial – itself play a role?

To answer these questions, we have to move to a higher level of analysis and examine the three ways in which the *local government* contributed to the events of May 2000. The Walkerton Public Utilities Commission is run by an elected group of commissioners who are legally responsible for its control and management. Nevertheless, they didn't concern themselves with the details of system operation, but chose to focus instead on budgeting and financial matters. In addition, they had very little knowledge of water safety or of principles of waterworks operation, and so relied almost exclusively on the senior management, primarily Stan Koebel, to identify and resolve any concerns related to the system. For example, when a 1998 ministry inspection uncovered serious problems, including the presence of E. coli in treated water samples, the commissioners simply accepted Stan Koebel's assurances that he would look after the situation. They didn't ask how the problems had arisen, nor did they follow up to ensure that Koebel actually addressed the concerns. As became clear later, Stan failed to follow through on his assurances. The inquiry following the disaster concluded that it was reasonable to have expected the commissioners to take a more active role in responding to the 1998 ministry inspection report.

The local government also played a role in the outbreak because of its weak role in publicizing the "boil water" advisory issued by the local health unit on May 21. It's estimated that only about half of Walkerton's residents actually became aware of the advisory on May 21. Some members of the public actually continued drinking from the town water supply until as late as May 23. The inquiry report suggested that this delay was partially attributable to the failure of Walkerton's mayor, David Thomson, to take an active role in building public awareness of the advisory after he was informed of it by Dr. Murray McQuigge, the local medical officer of health. The mayor was not specifically asked by Dr. McQuigge to assist in publicizing the advisory, and there is some dispute as to whether the seriousness of the situation was effectively communicated to him. In any case, there was no clear, pre-defined role for the mayor, and by extension the municipality, in ensuring that the public knew about any "boil water" advisory issued by the health unit. As a result, the mayor took no action and relied on the doctor to handle the situation. The inquiry report didn't place responsibility on the mayor, but observed that his inaction did represent a missed opportunity to reduce the scope of the outbreak.

These additional pieces of the puzzle interacted with the previously identified pieces – revealing the multiplicity of factors that contributed to the outbreak: not only does management matter in maintaining the safety of a complex technological system, so does government. But the chain of command and the responsibility curve didn't end there.

Ontario's Ministry of the Environment is the main regulatory agency with responsibility for overseeing the operation of municipal water systems in the province, and its actions – or more precisely, its inaction – contributed to the Walkerton outbreak in six ways. First, because of the ministry's subpar operator training and certification programs, Stan and Frank Koebel had a woefully inadequate understanding of water safety issues, including the public health risks posed by E. coli, and believed that untreated water from the wells was essentially safe for human consumption. Had they understood that unchlorinated water could kill, their behaviour would undoubtedly have been very different.

The weak response of the ministry to evidence of repeated violations – violations uncovered by its own periodic inspections – was the second significant factor that contributed to the substandard operating practices at the WPUC. Inspections conducted in 1991, 1995 and 1998 repeatedly revealed worrisome deficiencies in treatment and monitoring at the WPUC. After each inspection, the ministry made recommendations to address the deficiencies, but each and every time Stan Koebel either put off or ignored those recommendations. But no measures were ever taken by the ministry to ensure that their safety concerns were addressed, even though they had the legal authority to implement tough measures. Instead, the ministry relied on Stan Koebel and his staff's voluntary compliance with its recommendations. This soft regulatory stance only reinforced Koebel's belief that the recommendations and guidelines weren't important for maintaining the quality of the water.

Thirdly, the ministry didn't have access to the information that was required to monitor compliance with its regulations by local governments, not just in Walkerton, but across Ontario. As Commissioner O'Connor observed:

> The [ministry] did not have an information system that made critical information about the history of vulnerable water sources, like Well 5, accessible to those responsible for ensuring that proper treatment and monitoring were taking place. On

several occasions in the 1990s, having had access to this information would have enabled ministry personnel to be fully
informed in making decisions about current circumstances and
the proper actions to be taken.[17]

This lack of feedback was responsible for two other factors
that contributed to the outbreak. One of those was related to the
ministry's approvals program. As I mentioned, Walkerton's Well
5 didn't have continuous chlorine residual monitors installed.
When Well 5 was initially approved by the ministry in 1979, it
was identified as being susceptible to contamination from surface water. However, as was consistent with the ministry's practices at the time, no special conditions for monitoring were
attached to the approval. By the 1990s, special conditions for
monitoring and treatment were routinely attached to approvals
for wells similar to Well 5. Yet the ministry didn't attempt to
apply such conditions retroactively to previously granted
approvals, partly because it didn't have an integrated information system that would have allowed it to track older certificates
of approval.

The ministry's inspections program also contributed to the
tragedy. Although in 1994 a provincial guideline was amended
to require continuous, rather than the previous daily, monitoring of chlorine in wells at risk of surface contamination, no
attempt was made to systematically review existing certificates
of approval or notify water system operators of the amendment,
or to assess existing wells during inspections. In fact, there were
no criteria available from the ministry to guide inspectors in
determining whether or not a given well was at risk. The ministry inspections in 1991, 1995 and 1998 didn't detect the vulnerability of Well 5, even though the relevant information was
available in ministry records. But the records were archived and
difficult to find. So the continuous monitors that might have
prevented the outbreak were never installed for Well 5.

There is one other very important regulatory factor that contributed to the Walkerton outbreak. In 1996, laboratory testing of drinking water quality was privatized by the right-wing Conservative provincial government. Municipalities like Walkerton could no longer rely on government-run facilities and had to start using private laboratories. No legislation, however, was enacted requiring private laboratories to notify the ministry of adverse test results. Municipalities were expected to deal with any problems that arose. A number of people were aware of the risks this introduced for public health, but despite high-level discussions between the Ministries of Health and the Environment, the government hadn't acted to close this loophole. So on May 17, when A&L discovered that the water samples collected on May 15 were severely contaminated, only Stan Koebel was notified of the results, and so the community's safety depended on one incompetent, frightened man. The weak notification protocol did not require him to pass the information on, and he did not. The delays in discovering the source of the outbreak seriously exacerbated its impact.

The scapegoat explanation now has no credibility as the sole explanation for the outbreak. Yes, Stan Koebel lied to cover up his mistakes, and the WPUC didn't do what they were supposed to do. There is no excuse for that. But someone should have made sure that the responsible parties had the proper training, equipment and oversight to safeguard the quality of the water. That's why government entities like the Ministry of the Environment exist – to serve as a benevolent "invisible hand" that protects the public interest, particularly when the cost of failure in a technological system is lethal. To allow someone who doesn't even know what E. coli is, let alone that it's lethal, to be in charge of managing a public drinking water system and expect him to do his job well is tantamount to asking him to be superhuman.

But before we place all the remaining responsibility on the ministry, we would do well to ask if there are any systematic reasons that it did such a poor job at protecting the health of the Walkerton residents. To answer this question, we have to move to the final pieces of the puzzle, which reveal most clearly of all why political considerations are so critical to the safety of complex technological systems.

The shortcomings of the ministry were due, at least in part, to the "system design" decisions taken by the *provincial government* in the years leading up to the Walkerton tragedy. The Ontario Conservative Party's "Common Sense Revolution," based on a platform of "smaller government" and "fiscal responsibility," had appealed to prevailing public opinion, so voters gave it overwhelming support on election day. Premier Mike Harris's government then proceeded to slash the ministry's budget by nearly half. In the two-year period between 1996 and 1998 alone, the budget was cut by over $200 million, with a consequent staff reduction of over 30 per cent (more than 750 employees). Because of the ongoing budget cuts and staff downsizing, the ability of the "leaner, meaner" ministry to take a proactive role in detecting and preventing problems was systematically eroded. A former deputy minister testified at the inquiry that proactive inspections and follow-ups took on reduced priority because "the day was eaten up with reactive work."[18] These changes had a direct impact on the Walkerton municipality. Between 1994–95 and 1999–2000, the number of planned inspections by the ministry office responsible for Walkerton fell by 60 per cent, and the amount of employee resources dedicated to communal water decreased by almost half.

The potential harm to the public arising from these reductions was known, yet, according to Commissioner O'Connor, "Despite having knowledge that there could be risks, no member of Cabinet or other public servant directed that a risk assessment and management plan be conducted to determine the extent of those risks, whether the risks should be assumed, and if assumed, whether they could be managed."[19] Even without any risk assessment, "the Cabinet approved the budget reductions in the face of warnings of increased risk to the environment and human health."[20] These budget cuts substantially reduced the likelihood that the ministry's approvals and inspections programs would uncover the need for continuous monitors or the improper operating practices at Walkerton and elsewhere.

The second major role played by the provincial government concerned its "distaste for regulation" and its resulting decision to privatize laboratory testing of drinking water.[21] The failure of the government to enact legislation requiring private labs to notify the Ministry of the Environment and health authorities of adverse test results eventually contributed to the tragic events at Walkerton. The evidence presented at the inquiry clearly showed that high levels of government were aware that the lack of such legislation posed a potential risk. However, by 1995–96, the newly elected government's efforts to reduce regulation were in full swing: it created a "Red Tape Commission" to eliminate "complicated and unnecessary paperwork" resulting from rules such as reporting requirements. For similar ideological reasons, the government didn't act to require mandatory accreditation of private testing labs. The opinion of ministry officials given at the inquiry was that any move to legislate a notification requirement or accreditation would "likely have been 'a non-starter,' given the government's focus on minimizing regulation."[22]

The policy and regulatory decisions made by the provincial government interacted with all of the other contributing factors – another textbook demonstration of a systems accident – fully unravelling the mystery of why seven people died and 2,300 became ill in Walkerton, Ontario, during and after May 2000. The whole story offers an astonishing picture of the role politics plays in safeguarding or sabotaging complex sociotechnical systems. The "system design" decisions made at the highest levels of a government – budget allocations, regulations and policy goals – interacted with environmental factors (heavy rainfall, the

nearby manure and the geology surrounding Well 5), psychological factors (the complacency and deceptions of the Koebel
brothers, to name just a few), and organizational factors (the
incompetence and negligence of the local government and the
Walkerton Public Utilities Commission), creating a situation
where the ministry was crippled, thereby virtually guaranteeing
that it would eventually fail in its role as the province-wide overseer and enforcer of drinking water quality. Rather than acting to
safeguard public health – a compassionate "invisible hand" – the
government's decisions pushed an already flawed system beyond
the brink of disaster – a callous and careless "invisible hand."

The Ontario Conservative Party, with its electoral success,
not surprisingly believed that its platform of fiscal reduction was
tailored to its constituents' wishes. From this perspective, it
might seem as if the policies and regulations that resulted from
the "Common Sense Revolution" were the epitome of Human-
tech thinking. Yet the results of these political system design
decisions were catastrophic – they certainly didn't lead to an
improvement in quality of life. The Walkerton case study might
suggest that people died *despite* the fact that there was an affinity between the government policies and the prevailing political
climate in Ontario in the mid- to late 1990s. But while developing policies that are tailored to public opinion is necessary at
the political level to get into power, getting elected by making the
masses happy isn't enough. "System design" decisions made by
governments must also reflect affinity with human nature at
every level of the system – the organizational, team, psychological and physical levels.

From this broader perspective, the Ontario government's
decisions were a resounding failure: the new design of the
Ministry of the Environment didn't provide nearly enough
financial or human resources to safeguard the public interest;
the design of the local government organization didn't provide nearly enough expert oversight of the Walkerton Public

Utility Commission; the design of the WPUC management structures didn't provide nearly enough expertise to monitor the water supply system effectively; and so on, and so on. Under these conditions, it would have taken a superhuman effort at the lower levels to make the government decisions work well. At the ministry alone, public servants struggled to cope with the reduced human and financial resources at their disposal. The provincial government's "system design" decisions repeatedly went against the grain of the human factor, emphasizing, once again, that Human-tech thinking is not "common sense."

As long ago as 1759, our friend Adam Smith had the foresight to anticipate this pitfall as well as the resulting devastating damages in his *Theory of Moral Sentiments.* The arrogant politician of Smith's day was already

> . . . often so enamoured with the supposed beauty of his own ideal plan of government, that he cannot suffer the smallest deviation from any part of it. He goes on to establish it completely in all its parts without any regard either to the great interests, or to the strong prejudices which may oppose it. He seems to imagine that he can arrange the different members of a great society with as much ease as the hand arranges the different pieces upon a chess-board. He does not consider that the pieces upon the chess-board have no other principle of motion besides that which the hand impresses upon them; but that, in the great chess-board of human society, every single piece has a principle of motion of its own, altogether different from that which the legislature might chuse [sic] to impress upon it. If those two principles coincide and act in the same direction, the game of human society will go on easily and harmoniously, and is very likely to be happy and successful. If they are opposite or different, the game will go on miserably, and the society must be at all times in the highest degree of disorder.[23]

The sheer will and power of politics alone can't *force* organizations, teams or physical laws any more than it can force individuals to do things that are simply beyond their limited means. Each of these also has "principles of motion" that must be respected if the system as a whole is to be designed effectively, and a tight Human-tech fit must be achieved at *every* level of human nature, not just the political. We should remember (to paraphrase the words of the Nobel laureate Richard Feynman, after the *Challenger* space shuttle disaster) that for a technology to be successful, reality must take precedence over political ideology, because human nature cannot be fooled. This lesson is perhaps the most important legacy of the tragic Walkerton E. coli outbreak.

And the players? Stan Koebel eventually sold his house and left Walkerton for good. Frank Koebel was put on sick leave while lawyers tried to figure out a permanent solution for his job situation. The WPUC plugged up and abandoned Wells 5 and 6. The Ministry of the Environment said it would implement all of the recommendations from the Walkerton Inquiry Report, but it remains to be seen how many of the changes will take hold. Premier Mike Harris resigned, citing personal reasons, while polls showed that voter support for the provincial government plummeted to the point where it was 25 per cent behind the opposition party. A provincial election has yet to be held, so the Conservative Party is still in power. Many Walkerton residents still use bottled water to this day.

Just as this book is going to press, the Ontario Provincial Police have announced twelve criminal charges against the Koebel brothers, including public endangerment, forgery and breach of trust, which have a maximum penalty of up to ten years in jail. The Koebel brothers should be held accountable in some way because they knowingly violated provincial regulations, falsified records, withheld life-critical information and lied to cover up their incompetence. But one of the fundamental lessons of Human-tech thinking is that accidents such as Walkerton are systems accidents with multiple levels of accountability. Yet the provincial police are not planning to charge anyone else, not in the local government, not in the Ministry of the Environment, not even in the provincial government, despite the fact that the former premier and his cabinet had available highly educated advisors who knew what E. coli was and that it was lethal. These advisors had warned of the risks that severe budget cuts could have on public health—warnings that were wilfully and knowingly ignored.

Focusing on the Koebel brothers proves that the Ontario judicial system isn't prepared to absorb the fundamental Human-tech lesson of the Walkerton outbreak, a fact that should concern all of us because accidents of this type can only be prevented by redesigning the system as a whole, not by merely finding lowly scapegoats to punish.

MANAGING RISK IN A DYNAMIC SOCIETY:
THE RASMUSSEN FRAMEWORK

How do we prevent such accidents from happening again? At a descriptive level, learning what happened in Walkerton helps us understand how the safety of the public drinking water supply in a small Canadian town was violated, but chances are that that particular set of events will never be exactly duplicated, in Walkerton or elsewhere. So if a Human-tech approach is to prevent this kind of accident, we have to peel away the idiosyncratic details of Walkerton to get to the essential factors that threaten the safety of complex technological systems in general.

The groundbreaking work of Professor Jens Rasmussen, a Danish engineer, gives us the conceptual tools to make this transition from descriptive understanding to prescriptive intervention.

I was lucky enough to spend a year in Denmark working with Professor Rasmussen between my master's and my Ph.D. – a year that influenced my thinking tremendously – so I'm intimately familiar with his work. Since that magical year, Jens has been a generous and kind mentor to me, and we wound up co-authoring some influential papers together.[24] He has a precious, rare gift: he can see structure where others see only chaos. And on top of that, he has a tremendously valuable set of experiences, having spent forty years trying to understand how risk in complex technological systems can be better managed, first in the nuclear industry and subsequently in other sectors.[26] He radiates a Yoda-like wisdom that is unparalleled in my experience. Jim Reason – a British psychologist who dedicated his widely cited 1990 book on human error to Jens[27] – once told me that Jens was like an onion: every time you think you've learned everything you can from him, there's always another layer to

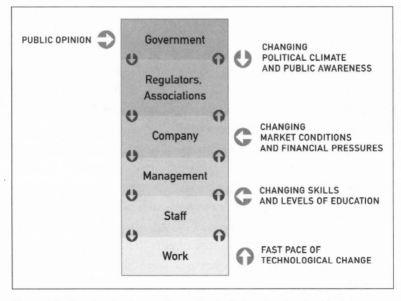

Figure 1. Various levels of a complex technological system involved in risk management.[25]

peel away. The crowning achievement of Jens's life work is a two-part, qualitative framework that aims to explain both how accidents occur and how they can be prevented.[28]

The first part of the framework, illustrated in figure 1, shows that virtually all complex technological systems comprise a hierarchy of players – both individuals and organizations. The exact number of levels in the hierarchy and the labels for them will differ across various industries, but the illustration shows the idealized, generic structure that these hierarchies tend to exhibit. The bottom level describes the technical and environmental factors associated with the particular (potentially hazardous) work (e.g., a water supply system, a nuclear power plant, commercial aviation). The next level describes the activities of the individual staff members who are responsible for doing the front-line work (e.g., water quality inspectors, control room operators, airplane pilots). The third level from the bottom describes the activities of the mangers who supervise the staff. The next level up describes the activities of the company as a whole. The fifth level describes the activities of the regulators or associations that are responsible for constraining the activities of companies in that particular sector. Finally, the top level describes the activities of government, both civil servants and elected officials, who are responsible for setting public policy. These political activities are, in turn, firmly shaped by public opinion.

Decisions at higher levels should propagate down the hierarchy, whereas information about the current state of affairs should propagate up the hierarchy. These interdependencies across levels are critical to the successful functioning of the system as a whole. If instructions from above aren't formulated or carried out, or if information from below isn't collected or conveyed, then the system can become unstable and start to lose control of the hazardous process that it's intended to safeguard.

Looked at in another way, safety can be viewed as an emergent property of a complex technological system. It's affected by the decisions of *all* of the players – politicians, CEOs, managers, safety officers and work planners – not just the front-line workers. Consequently, accidents or threats to safety usually result from a loss of control caused by a lack of vertical alignment – mismatches – across levels of a complex sociotechnical system, not just from deficiencies at any one level alone. Walkerton is no exception; all levels – from the environmental to the political – played a critical, albeit different, role in maintaining safety. As a result, it becomes clear that threats to safety or accidents are usually caused by multiple contributing factors – relationships – not just a single catastrophic decision or action. Understanding this is immensely important.

In turn, the lack of vertical alignment is frequently caused, in part, by a lack of feedback across levels. People at each level can't see how their decisions interact with those made by people at other levels, so the threats to safety aren't obvious until an accident occurs. In the Walkerton outbreak, the Walkerton Public Utilities Commission didn't receive feedback about the true quality of the drinking water because they frequently failed to take samples; the local government didn't receive enough feedback about the performance of the WPUC; because they trusted Stan Koebel; the Ministry of the Environment didn't receive enough feedback about the municipalities it was supposed to be overseeing because it didn't have an adequate information system; and the government didn't receive enough feedback about the impact that its policy decisions would have because it didn't conduct any risk analysis. Each level in the hierarchy was operating almost independently of all of the other levels. This was a system in the throes of chaos – a virtual recipe for disaster.

To understand how accidents happen in complex systems – death by interactions – we also have to realize that the operation of each of the levels in Jens's framework isn't written in stone. On the contrary, each level is constantly influenced by societal forces that stress the system as a whole. Examples of such interpenetrating stressors include changing political climate and public awareness (the "Common Sense Revolution"), changing market conditions and financial pressures (the ministry budget cuts), changing competencies and levels of education (more stringent qualifications to operate a public drinking water system), and changes in technological complexity (new equipment and more sophisticated tests to monitor water quality). In our dynamic twenty-first-century society, these external forces are stronger and change more frequently than ever before.

The net result of these stressors is captured by the second element of Jens's framework, shown in figure 2, which summarizes the dynamics generated by two monumental "invisible hands" – intangible but extremely powerful influences that can cause the structure and behaviour of a complex technological system to change over time. The large arrow pointing downward represents the financial pressures that push the people in the system to work in a more fiscally responsible manner. The ministry's budget cuts are an example of this "cost gradient." The large arrow pointing upward represents the psychological pressures that push the people in the system to work in a more mentally or physically efficient manner. The ways in which the Koebel brothers cut corners while doing their job provides an extreme example of this "effort gradient."

As a result of these two gradients, work practices are subject to an exploratory but systematic migration over time, represented by the crooked line moving from right to left across figure 2. Just as the force of gravity causes a stream of water to flow inevitably down the crevices in the side of a mountain, these financial and psychological forces inevitably cause people to try out different ways of working to find the most economic ways of performing their job. Moreover, the change in how people do their jobs can occur at multiple levels of a complex sociotechnical system as shown in figure 1, not just one level alone.

Over time, this migration causes people to cross the official boundary of work practices – the way you're supposed to do your job, "by the book." People are forced or choose to deviate from procedures, cutting corners because they're responding to requests or demands to be more cost-effective. Mind you, complex technological systems are usually designed with several layers of safeguards, an approach known as "defence in depth." However, the migration of work practices can cause these defences to degrade and gradually erode over time, not all at once. For example, downsizing at the ministry continued over a number of years, and the deviation in work practices at the WPUC had been going on for twenty years.

We might think that this lack of procedural compliance and this degradation in safety would raise an immediate red flag, but there are two reasons why they don't. First, people have to modify the way they do their work to get the job done, given the stresses that any complex system is undergoing. That's why

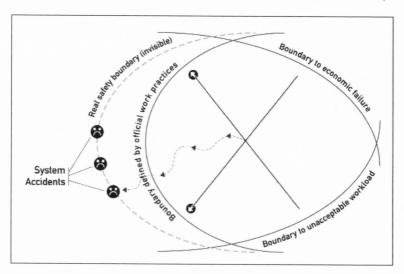

Figure 2. A model showing how financial and psychological forces can create behaviour gradients that cause work practices to migrate systematically toward the boundary of safety. Adapted from Rasmussen (1997).

"work to rule" campaigns, where people do their job *strictly* by the book, usually cause complex technological systems to come to a grinding halt. Second, the change in work practices doesn't usually have any immediate, visible, negative impact. The threat to safety isn't obvious before an accident because the violation of procedures doesn't immediately lead to catastrophe. At each level in the hierarchy, people are working hard, striving to respond to cost-effectiveness measures, but they don't see how their decisions interact with those made by other people at different levels of the system. Yet the sum total of these uncoordinated attempts at adapting to societal stressors is slowly but surely pushing the system and "preparing the stage for an accident," to use Jens's apt phrase.[29]

In the case of Walkerton, the changes unleashed by the "Common Sense Revolution" and the migration in practices that had been going on for years at the ministry and WPUC had no immediate visible impact on safety. Nobody died or suffered any obvious ill effects from drinking the water and that gave all concerned a false sense of security. Yet in retrospect we can see that the system as a whole was slowly but surely inching toward disaster. But since nobody in such situations can see that at the time, more budget cuts are introduced, the need for improving efficiency increases, and the change in work practices continues. People try harder to work in more efficient ways, and with each innovation they move closer and closer to the real boundary of safety on the far left of figure 2. But because that boundary is usually invisible because of a lack of feedback, people don't have any idea whether the technological system as a whole is moving toward or away from disaster. Deviations from official work practices can persist and evolve for years without any breach of safety, until at last the real safety boundary is reached. Some of the factors contributing to the Walkerton E. coli outbreak were so deeply entrenched that they had been habitual for more than twenty years. After an accident, workers will wonder what happened, because they just did things the way they'd "always" done them, at least in the recent past. In other words, accidents in complex technological systems don't usually occur because of an unusual action or an entirely new, one-time threat to safety. Instead, they result from a systematically induced migration in work practices combined with an odd event or coincidence that winds up revealing the degradation in safety that has been steadily increasing all the while. A systems accident is a death by interactions that are out of control because they haven't been designed to conform to human nature at all levels, nor to withstand the turbulence endemic to modern times.

From the perspective of Jens Rasmussen's rich framework, we can now see that the events in Walkerton could happen not only in other towns, but also in complex technological systems in other sectors, like space travel, nuclear power, or health care – the SARS epidemic and the West Nile virus come to mind as timely examples. Accidents of this type are a byproduct of the life we are currently living, in a world filled with complex technological systems, fallible human beings and dynamic societal pressures. However, if we take certain precautions, disasters need not occur. And therein lies the value of Rasmussen's framework: it shows how the Human-tech approach can be used to design complex technological systems that fulfill important societal needs while simultaneously safeguarding the public interest.

Adding more defences-in-depth alone is unlikely to work in the long run because the new defences will just delay the inevitable degradation that's caused by the financial and psychological gradients shown in figure 2. Creating more awareness through public relations or educational campaigns alone is also unlikely to be a viable long-term solution because the same two "invisible hands" will always be there, prodding people to migrate to more cost-effective and mentally and physically efficient ways of doing business.

The only long-term solution to managing risk in a dynamic society like our own appears to involve first of all accepting that external stressors such as budget cuts and market competitiveness aren't going to go away entirely since they're the result of persistent human factors. Then we can focus on deliberately building technological systems that can respond and adapt to these pressures without compromising safety. In other words, the goal is to allow systems to operate "at the edge" to maximize competitiveness or efficiency, but without actually breaking the envelope of safety and precipitating accidents. To "operate at the edge," vertical alignment via feedback across levels must be achieved so that each person and organization in the system can see the effect their actions have on safety, not just on the bottom line. There are five conditions to implementing this vision:

1 Identify the *players* – both individuals and organizations – that are involved in making decisions in the particular complex sociotechnical system under investigation. The hierarchy in figure 1 is a useful starting point for this step.

2 At each level of the hierarchy, the decision-makers should have explicit *work objectives* that they're responsible for achieving. Moreover, the objectives at the various levels should be synergistic, not conflicting.

3 *Feedback* should be provided at each level of the hierarchy so that decision-makers can assess the current state of the system and see how well it corresponds to the desired state defined by their respective work objectives.

4 Decision-makers at all levels should have the *capabilities and competencies* to interpret this feedback, compare it to their work objectives, and then decide what courses of action need to be taken, if any.

5 All the players in the system should have the required *commitment* to achieve their respective work objectives so that safety can remain a priority in the face of other competing issues.

A complex technological system that's designed according to these conditions will respect human nature through and through – all the way from the political level to the physical level, resulting in one harmonious system, a giant, benevolent "invisible hand" that will make tragedies such as Walkerton less likely to become commonplace occurrences, as the winds of change continue to blow in our dynamic society.

A public water distribution system can produce drinking water or poison. The hardware to achieve either goal is roughly the same. This just goes to show us how truly insignificant the strictly technical components are by themselves. Coming as it does from an engineer, this statement might seem heretical, particularly in the twenty-first century where technology reigns supreme. But I stand by my claim. The real power isn't in software or hardware alone, but in how people use it. Sure, technology is an enabler that can create new and wonderful possibilities, but if we can take the very same system and either poison or sustain people with it, then the inescapable conclusion is that all of the interactions involving the human factor are incredibly important influences on the safety of any technological system. Therefore, the true promise of the Technological Age isn't just in the things that we build, but in how we use those things as tools to help improve our quality of life. *That* is one of the fundamental lessons of the Human-tech approach.

Unfortunately, not everyone thinks this way – yet.

Part Three

REGAINING CONTROL
OF OUR LIVES

10

The Way Forward: Not by Widgets Alone

On Tuesday November 7, 2000, over 100 million Americans tried to participate in the democratic process. The outcome of the presidential election would affect the future of their country for the next four years. Would it be the Republican candidate George W. Bush, the Democrat Al Gore, the Green Party nominee Ralph Nader or Pat Buchanan (who ran as an independent)? Eventually, everything boiled down to the outcome in Florida. Whoever won that state would become the forty-third president of the United States of America. The *last* final tally (there were several) showed that Bush beat out Gore in Florida, but only by the slimmest of margins – 2,912,790 votes compared to 2,912,253; the difference was a mere 537 votes, or 0.0092 per cent of the total vote in Florida.[1] If the Florida contest had been a marathon race run at world-record pace, then Bush would have beaten Gore by a split second. But as many of us learned after a month of media coverage about hanging chads, dimpled chads, pregnant chads, hand recounts, machine recounts, butterfly ballots and Supreme Court decisions, this was not just your run-of-the-mill close election.

Many voters in Palm Beach County, Florida, were confused by the design of the so-called "butterfly ballot" and mistakenly voted for right-wing conservative Patrick Buchanan, instead of liberal democrat Gore – as if there was an invisible (right) hand influencing their actions. This design-induced "human error" led Buchanan to garner 3,407 votes in the staunchly Democratic county. That's 2,400 more votes than he received in any of Florida's other sixty-six counties – a suspiciously good showing considering that Buchanan never made a single campaign stop in Palm Beach County. In an interview with NBC, Buchanan himself admitted that the ballot layout was confusing and that many of the votes he received were meant for Gore.[2] A scientific experiment conducted by psychologists at the University of Alberta later confirmed Buchanan's intuitions; changing the ballot design did indeed significantly reduce the number of voter errors.[3]

The problems with the ballot design in Palm Beach County also led to a large number of spoiled votes; 19,120 presidential ballots weren't counted because they showed more than one vote. In comparison, only 3,783 senatorial ballots were double-punched in the same county.[4] Many Gore supporters said they mistakenly voted for Buchanan on their first attempt, realized their mistake, and then punched a second vote for Gore. Perhaps they believed they could correct their mistake, but ballots punched twice weren't counted – not even once. As a result, those citizens weren't able to exercise their democratic right to vote – an international embarrassment for a nation that prides itself on individual freedom and liberty.

The consequences of the confusing design were extraordinarily important in this case. The arithmetic is crystal clear. Were it not for the thousands of mistaken and spoiled votes, the liberal democrat Gore, not the right-wing conservative Bush, would have been president of the country for four years. The integrity of the democratic process was violated. As one ensuing court order put it, "The ballot was designed and printed in such a way that voters were deprived of their right to freely express their will."[5] A lack of affinity between people and technology changed the course of world history – a fact that should give us pause, in the context of recent world events.

But the reaction to the Butterfly Ballot debacle is at least as revealing as the events themselves. Many people blamed the "voter errors" on individual frailty rather than on inadequate design. The Palm Beach County commissioner said "The ballot is very straightforward. You follow the arrow, you punch the location. Then you have voted for who you intend to elect."[6] Thousands of people in Palm Beach County disagreed.

Roll the calendar forward six months, and we see an even more revealing response. On May 17, 2001, the Associated Press reported that politicians in Palm Beach County, Florida, allocated $16 million to replace punch-card voting machines with a computerized "touch-screen" system that purported to be the most advanced and was certainly the most expensive method available.[7] That seemed like a smart public policy decision made possible by technological progress. Although the financial investment would be large, it could prevent an embarrassing voting fiasco from ever occurring again. Computerized voting, it was assumed, could fix all of those problems with hanging chads, dimpled chads, pregnant chads, hand recounts, machine recounts and butterfly ballots.

But the thinking was traditional: Mechanistic stuff. Not Human-tech.

Shortly after the presidential election, scientists at MIT and Cal Tech began researching the reliability of different voting technologies.[8] They looked at five alternatives: manually counted paper ballots, mechanical lever machines, paper ballots that are optically scanned by machines, the punch cards that had been used in Palm Beach County, and the touch-screen-based electronic voting systems that Palm Beach County officials wanted to invest in after the election. Using historical data from elections held between 1988 and 2000, the scientists identified the average percentage of uncounted, spoiled or unmarked ballots generated by each type of technology.

The results would surprise anyone who thinks that the voter errors observed in U.S. Election 2000 can be reduced by throwing fancy technical expertise at the problem. The computerized touch-screen-based systems produce one of the *highest* ballot error rates of any voting technology. The only good thing about this "advanced" technology is that it isn't appreciably worse than the notoriously problematic punch cards that were already being used.

From a Human-tech perspective, one of the problems with the touch-screen is that it doesn't take into account voters' psychological habits and expectations. Some people, when they read text on a computer, will put a finger on the screen and move it over the text. Performing that innocent action has no effect on a paper ballot, but with the unfamiliar touch-screen device it could unintentionally record a vote for the first name that's touched, and the erroneous vote may be hard to change. One of the researchers who conducted the study pointed out the likely result: "the voter gets frustrated, cancels the session, and walks out."[9] High-tech is not the same as Human-tech.

Which voting technology was the most reliable? The optically scanned paper cards – a technology that had been around for decades – were best, followed by manually counted paper ballots – the oldest of the currently used voting technologies. Good old-fashioned paper beats out twenty-first-century computer technology – low-tech can be Human-tech. Even the eminent Cal Tech and MIT scientists who conducted the study were astonished by the results they obtained: "We were very surprised by the relatively high [error] rate of electronic equipment. When we began this investigation we expected the newer technologies to outperform the older technologies"; "paper ballots do remarkably well."[10] Nevertheless, the scientists pinpointed the

Human-tech implications of their results: "The [voting machines] industry needs to study user interface designs carefully: design and perform experiments using actual [sic] people in order to optimize the usability of the interfaces."[11]

Despite the extremely embarrassing and internationally publicized disaster with their voting systems and the availability of the Cal Tech/MIT research, the Palm Beach County officials went ahead with the touch-screen technology, incapable of seeing the fundamental lesson of U.S. Election 2000: to have a reliable voting system, the technology must be designed to have affinity with human nature. The Florida butterfly ballots should have been replaced by a technology that everyone can actually use rather than by a new, "advanced," expensive and equally unreliable alternative.

THE TIME HAS COME:
TRANSITIONAL INSTABILITY BUSTING AT THE SEAMS

I believe that we're currently at a critical crossroads in the history of civilization. Most people are ready for a new way of thinking about the role that technology should play in society. Individually, we've had enough of the turbulent, frustrating effects inflicted by the traditional current of thought, although of course, we don't want to get rid of technical innovation because we all know it has made changes for the better in our lives and holds the promise of more. But it can also have terrible effects: and they seem to be getting worse. We're all sick and tired of witnessing technology-induced death and destruction, like the tragic events at Chernobyl and Walkerton. The moral and financial costs of these calamities are simply unacceptable. We want to regain control of our lives, and we'd also like to see technological innovation used to improve, not threaten, the quality of life of everyone else on the planet. When we open our taps, we want clean drinking water, not poison. The pre-conditions for change are all around us.

I also fiercely believe that a Human-tech Revolution will help get us out of the detour of "transitional instability" that we're currently mired in, and back onto the long, winding, but inevitable road of cultural progress. Human-tech thinking embraces a more sophisticated social structure than the antiquated Mechanistic and Humanistic approaches because it insists on fully bringing together knowledge of both human nature and the physical world, not just one or the other in isolation, or the two in imbalance. Using such a more mature form of coordination, we can harness and exploit the increasing complexity of the technological systems that surround us. And it's not just a matter of wishful thinking: I think it's inevitable. Just as agricultural chiefdoms (or something like them) were bound to appear to make full use of farming technology, Human-tech thinking (or something like it) is bound to rise in prominence to harness and exploit the complexities of twenty-first-century technology.

Adopting a new way of thinking *is* inevitable when it's fuelled by need. And the relationship between technology and people is now so entrenched that, directly or indirectly, it affects the daily activities of everyone on the planet.

Human-tech isn't a household word yet, but it has already had a significant impact on our quality of life and some basic principles are in place to help bring it to the fore in practice. We already know how to design technology with a human face. Here's a summary of some of the examples we've covered; and you'll find they have wide application in many areas of life, business and industry:

- *Task analysis* is a method that can be used to understand the steps that are required to perform a particular job. This understanding allows us to design things that help people get their jobs done more easily – for example, to design a lathe that a normal human being can use; you wouldn't have to be four foot three inches tall, have shoulders that are two feet wide, and an arm span of eight feet to operate it comfortably.

- *Stimulus-response compatibility* is a design principle that creates an obvious relationship between a device's controls and the things being controlled. Products that are designed this way are easier for people to operate – such as a stove that eliminates human error, as opposed to one that leads you to burn a kettle every once in a while.

- The principle of *behaviour-shaping constraints* makes it easy for people to do the right thing and difficult or impossible for them to do the wrong thing. Products and systems that incorporate this principle make human error less likely, and desired actions more likely. For example, the Power Pig display on PCs was designed to make it very easy to conserve electricity, thereby encouraging environmentally responsible behaviour. And anaesthesia machines now make it physically impossible to connect a gas hose to the wrong nozzle.

- The *feedback* design principle makes the effects of our actions visible to us in a prompt, salient way. Products and systems designed using this principle make it easier for people to see how they need to change their behaviour to achieve their goals – such as the photocopier display designed by some of my clever undergraduate students, which would allow people to minimize the amount of paper wasted, contributing toward the goal of paper conservation.

- *Shape coding* is a design principle that makes it easy for people to distinguish between controls that do different things, even if those controls are right next to each other. Products designed using this principle are less likely to lead to errors. For example, during World War II, the flaps and the landing gear controls on military aircraft were redesigned so that they were easy to tell apart merely by touch, and as a result accidents caused by a lowering of the wheels after landing were eliminated.

- The *Aviation Safety Reporting System* (ASRS) – the incident reporting system that's geared toward learning from experience – allows people to share critical information without being penalized, and then provides a mechanism for sharing that information with others. The ASRS has been tremendously successful in improving aviation safety. It has saved lives. Variants of it are now beginning to be used in other industries, such as health care.

- *Cockpit Resource Management* (CRM) – the training program that takes advantage of simulator technology and psychological research on team dynamics – helps people work effectively together when dealing with complex technological systems. CRM training is now mandatory in the aviation industry and has led to significant improvements in safety – it too has saved lives. Happily, variants of it are beginning to be used in other industries, such as anaesthesiology.

- The *critical-incident technique* is a method for interviewing people to gather lessons learned from near misses or accidents. It can be used to identify problems in the relationship between people and technology, and suggest solutions to overcome those problems. As we've seen, it has been used by anaesthesiologists to greatly reduce the mortality rate associated with anaesthesia.

- Jens Rasmussen's *framework for risk management* shows how political and organizational decisions influence the safety of complex technological systems, like the Walkerton drinking water system. Systems design that results in a vertical alignment across the various levels of a socio-technical system can safeguard the public interest, even in the face of the stressors that typically threaten a dynamic society.

And that's not all. Even though it's still infrequently applied, this kind of thinking has led to a truckload of other design principles and achievements I haven't mentioned, all of them concentrated on ensuring that the human factor remains front and centre. And the impact hasn't just been on the industries that I've described in this book. It's also been used to make improvements in many other important social or industrial areas, such as air traffic control, products for the elderly and the disabled, and occupational health and safety, to name just three. Even though Mechanistic thinking still predominates in most sectors, what I'm advocating we call the Human-tech view has already had a positive impact on society. Just imagine the kinds of changes that could be unleashed if designing technology to have an affinity with human nature were a routine and widespread way of doing business. The world would be a completely different place, and certainly a better place.

Actually, we won't have to merely imagine what a Human-tech world would be like because, if my extension of Robert Wright's views on the logic of human destiny is correct, then this radical shift in thinking *will* happen. It's begun and we can't stop it. In aviation and nuclear power, Human-tech revolutions are well underway and we can see why; the move makes so much sense. It's clearly critical in terms of the safety of mass numbers of people, given the horrendous catastrophes we've experienced. The only uncertainty is *when* – when we will move from a state of blindly competitive technological chaos to a new, more humane equilibrium, in which the human factor is recognized as vital and technology is adapted to people rather than the other way around. Given the social turbulence we've experienced in the Information Age, and the signs of transitional instability all around us, I don't think we'll have to wait long.

The big missing step is to collect the dispersed strands of progress and bring them into focus to expose the general pattern.

By providing a label, a vision, and a conceptual framework, I hope to bring Human-tech thinking front and centre in people's consciousness to ensure that it gets the full attention it deserves.

WHAT CAN YOU DO TO MAKE A DIFFERENCE?
A BLUEPRINT FOR CHANGE

There's no sense in complaining about a problem unless you've got some ideas about how to fix it. The Human-tech Revolution has to advance concurrently on multiple levels. All of us – citizens, corporations, governments, international human development organizations, media and universities – have an important role to play in making the transition from the traditional currents of thought – the Mechanistic and Humanistic world views – to an enlightened and liberating recognition of the importance of the human factor in technological systems.

For *citizens,* one of the important points is to echo what Don Norman so effectively proposed in his best-selling book *The Psychology of Everyday Things:*

> If you are a user, then join your voice with those who cry for usable products. Write to manufacturers. Boycott unusable designs. Support good designs by purchasing them, even if it means going out of your way, even if it means spending a bit more. And voice your concerns to the stores that carry the products; manufacturers listen to their customers. . . . Walk around the world examining the details of design. Take pride in the little things that help; think kindly of the person who so thoughtfully put them in. Realize that even details matter, that the designer may have had to fight to include something helpful. Give mental prizes to those who practice good design: send flowers. Jeer those who don't: send weeds.[12]

In short, if you want to live in a world that celebrates human-ity and the human factor, then buy Human-tech products. Begin to distinguish and recognize poorly designed products. You'll stop blaming yourself for being technologically incompetent. Tell your friends about them. Show them how much better a Human-tech gadget is, such as a PalmPilot, than one dominated by 10 million features; your friends will thank you. They too will be more likely to buy products that have an affinity with human nature. And that, in turn, will make the Wizards listen. There's nothing like market pressure to encourage companies to change the way they do business. Human-tech consumers will eventu-ally drive out Mechanistic designs.

And what about our health care, or our safety when we fly, or the state of our environment? Most of us don't buy nuclear power plants, airplanes or hospitals. Can we still do something to improve the design of these complex, safety-critical sys-tems? I think we can. By actively participating in society as educated citizens, we can influence the democratic process to ensure that Mechanistic blindness doesn't dominate when important decisions affecting all of us are made. We can remind people–like the politicians in Palm Beach County who spent $16 million of tax payers' money on "advanced" voting machines – that ignoring the human factor has weighty impli-cations. Statistics based on solid scientific research can come in very handy in this respect.

Here's an example, based on my personal experience. When I tell people – as I told you at the beginning of this book – that between 44,000 and 98,000 preventable deaths occur each year in hospitals in the United States alone, and that this is equiva-lent to a wide-body jet aircraft accident with no survivors occurring once every day or two, people tend to react in pre-dictable ways. First, they stop what they're doing and listen. You've grabbed their attention, which is the first step toward learning. Second, they can't believe it's true. Then you introduce them to the study conducted by top-notch medical researchers at Harvard and published in the *New England Journal of Medicine*. The third reaction you'll get is surprise – why didn't I know that? Why don't more people know that? Good questions, of course. Once you get people this far, you'll find all kinds of things follow quickly. They'll want to know if there's a way to address the problem. You tell them that aviation has improved safety by paying more attention to Human-tech thinking than any other industry and that the same could be done for health care. They'll react by asking why more people aren't following that path, why are lives being lost unnecessarily? Once you get people this far, the move to a Human-tech world view has begun.

Executives and managers of *corporations* also have an essential role to play in furthering a Human-tech revolution. Contrary to popular belief, many people in positions of leadership in industry aren't driven exclusively by a desire for money and profit. They're also driven by vision and a passion to change society for the better – to improve quality of life.

There doesn't have to be a trade-off between market share and quality of life – the two *can* go hand in hand. But in most cases, a full recognition of this and what it might achieve requires profound changes in the way a company designs its systems as much as its products: people's needs will have to be put ahead of technology for its own sake; potential users of the product need to be consulted and involved from the very start; prototypes built and tested with real users (not Wizards) to see what works and doesn't work – and the results from these tests need to be iteratively used to improve the design of the product. Human-tech design principles need to be applied; and mechanisms should be put in place to get feedback from customers after they've made their purchase to get innovative and useful ideas for the next generation of products. A few companies that have already made these kinds of changes are reaping the benefits because designing technology that leverages the human factor makes good business sense. The Reach toothbrush, the PalmPilot, the Fender Stratocaster – these are all examples of products that were designed with a Human-tech approach, and each of these products has enjoyed tremendous success in the marketplace. Consumers reward companies for designing technology that honours the human factor. And the same kind of thinking could be applied to products that affect pressing social problems, like those associated with the environment.

If we really wanted to push the envelope, we could even use Human-tech thinking as a market differentiator. Imagine the following ad for a company that makes mechanical lathes: "Here's what our competition thinks you look like [accompanying image of midget extraterrestrial who's four foot three inches tall, has shoulders two feet wide, and an arm span of eight feet]. At Human-tech Inc., we design lathes for people. . . ."[13] We could create ads like that for all kinds of companies selling all kinds of products, and customers would instantly identify with the message because *they're* the ones who feel the effects of ignoring the human factor. A Human-tech advertising campaign could bring in a wealth of new business, and deservedly so for companies who design technology that works for people.

But it's not just product design companies that can benefit from Human-tech thinking. All but the smallest of companies require a well-designed bureaucratic infrastructure – a harmonious system – to succeed. But we have usually done such a poor job designing these teamwork and organizational processes that the word "bureaucracy" now has incredibly negative connotations, even though when you look it up in Webster's online dictionary, the first definition given is: "1 a: body of non-elective government officials; b: an administrative policy-making group."[14] That sounds useful, even necessary. How did we get from that definition to the kinds of inane situations described in the *Dilbert* comic strip? We've forgotten about the human factor, so we design corporate processes that no human being could possibly keep up with, let alone follow.

Often, corporate priorities are muddy or counterproductive – people who are meant to be working together instead may unknowingly pull in different directions, or even worse, actively compete against each other. Also, responsibilities are not explicit, or if they are, they may contain structural conflicts of interest (as when the FAA was responsible for incident reports and disciplining). And even if all of those difficulties are dealt with, the right information still doesn't always get to the right people at the right time; the communication patterns required to get the job done effectively and efficiently haven't been built into the team or organization. Even with the right information, individuals may not have the required expertise to interpret that information and use it to make effective decisions. Sometimes, people don't know what they don't know, so it's up to the "system designers" of the corporation to determine what skills are necessary for each job and to ensure that all employees have the requisite skill set.

And then there's the curse of procedures, rules, guidelines or policies that are supposed to shape work processes. I've been in workplaces where there are procedures for changing procedures! (I wonder how those got implemented? An act of God, perhaps.) And as if that wasn't bad enough, the rules are growing at an astronomical rate. René Amalberti, a French physician and psychologist and one of the leading Human-tech thinkers in Europe, noted that the number of rules, guidelines and policies in the European Joint Aviation Regulations is increasing at the rate of two hundred per year.[15] Is it reasonable to expect human beings to work effectively within the confines of an ever-tightening procedural straitjacket? Is it surprising that when things go wrong, investigations *always* reveal that someone violated a procedure?

Of course, some procedures are always necessary and must always be followed – we don't want a repeat of Walkerton. But often in cases where many rules are indeed required – as in very large organizations – people aren't given the resources to follow them, so they don't. I'm no exception. Once a year, I have to fill out a form to verify that I complied with all of the many policies and procedures in my university, and while I do my best to follow them, every year I write exactly the same thing at the bottom: "We have too many rules, and not enough resources to follow them. Things are bound to fall through the cracks."

Companies need to pay more attention to the human factor – to the people who are required to follow all of these rules and work processes. If companies took a Human-tech approach and looked at the system as a whole, they could figure out which administrative relationships really are necessary, and design "soft" systems so that the people in place have the competencies, information, goals and commitment to do their job effectively, without having to be supermen or superwomen. This isn't easy and it takes time and thought, but the kind of thinking that went into the development of Rasmussen's framework for risk management might be adapted to align the relationships and improve the harmony of corporate systems. We should be able to design bureaucracies in the true sense of the word: team and organizational structures that improve our quality of working life, and that help a company work smartly – and profitably.

Governments – both politicians and civil servants – can also play a critical role in a Human-tech Revolution. Our government officials generally don't have the greatest reputation in the public eye, but many people go into public service because they want to make the world a better place and are willing to work very hard, very passionately, to achieve that goal. Governments are of course responsible for regulating industries that have a strong technological component. But the complex nature of technology in modern times means that our political systems must also take the human factor into account at every level. By ensuring that Human-tech thinking is an integral part of regulations that govern safety-critical sectors, safety could be improved and lives saved. The aviation industry provides a role model for this type of change, which is one of the reasons why flying is safer than ever. New regulations in the nuclear industry in the United States have made plants safer than they were in the Mechanistic days before the Three Mile Island accident. The Walkerton disaster shouldn't have happened, but public water systems will – we hope – be better regulated in the future as a result. The same approach could be adopted in other industries. For instance, patient safety could be improved if Human-tech thinking played a more prominent role in medical regulations, by mandating that drugs be labelled clearly so that they're not easily confused with one another, for example.

In the case of nationalized corporations or services, governments could provide incentives to make the transition to the Human-tech age. In countries with publicly funded health care, for example – a list that includes Canada, the U.K., Denmark, the Netherlands, France and many others – governments could encourage hospitals to buy medical devices only from companies that followed a Human-tech design philosophy.[16] Mechanistic companies that didn't design technology with an affinity for human nature would quickly change their design practices.

For the most part, government hasn't appreciated the significant contribution that Human-tech thinking could make in helping address some of society's important problems. Yet progress on such issues will be difficult to achieve, and even more difficult to sustain, unless they have a higher priority in policy agendas. This too needs to change.

Governments can also adopt new legislation to facilitate some of the changes that are advancing the Human-tech Revolution. An outstanding example that is just begging to be implemented in many countries is the creation of national, non-punitive medical incident-reporting systems – the health care equivalent of the ASRS. Implementing such a system would require legal changes to protect the identity of the people reporting the incidents so that reported information would not be available in civil or criminal legal proceedings. Otherwise, a transition from a who's-to-blame system that hinders learning to a what's-to-blame system would be difficult to achieve. If the overwhelmingly positive experience of the ASRS in the United States is any indication, such changes would result in tremendous improvements in patient safety. Creating an effective mechanism for learning from experience could help end, for example, the chain of vincristine errors that have killed young children, even in countries with world-class health-care systems.

Government changes aren't easy to achieve because they involve coordination across a very large number of stakeholders. The first task will be mustering the required political will, and that alone will be a huge job. But people will make the effort if they understand that it can reduce the number of children, women, and men who die annually from preventable human error – currently as many as 98,000 people in U.S. hospitals alone, remember.

Both governmental and non-governmental *international human development organizations* also have a unique role in making the Human-tech Revolution a global reality, and addressing our most colossal societal need of all – closing the staggering and ever-widening gap between the rich and the poor. Most of the issues I've discussed in this book have centred on complex technologies in the so-called developed world, but the Human-tech view is critical in its implications for so-called developing countries. There are tremendous differences in culture and variations in social structure across countries – as a result, the technology that succeeds in one place may be a dismal failure in another. A Human-tech approach is urgently needed.

We can't expect complex technology to function effectively unless a country has social structures to receive and use it – or even know what to do with it. For instance, it probably wouldn't do much good to give a band of hunter-gatherers an Internet connection; they wouldn't know what to do with it. This might strike you as an absurd example, and it is, but consider the following real, yet equally absurd examples, reported by the Canadian journalist John Stackhouse in his book *Out of Poverty: And into Something More Comfortable*:[17]

- The German government donated a chemical reactor to a chemistry department in a university in Uganda, but the reactor didn't come with a manual or any instructions. As a result, nobody used it. In any case, what the department *really* needed wasn't a modern chemical reactor, but light bulbs.

- An international foreign aid agency provided funding for a hospital in Timbuktu, and sent some equipment with a dentist's chair and the technology to make false teeth. This too seemed like a good idea. Just two problems: Timbuktu didn't have any dental technicians, and there was no money to hire any. Once again, these obstacles probably didn't much matter. What the people of the Ivory Coast really needed was enough food to chew on.

- In India, a doctor wrote a prescription for a year's supply of birth control pills for a couple who already had nine children. Better late than never, you might think. Surprisingly, the couple used up all the pills in a mere six months. Why? Because both the husband and the wife had each been dutifully taking a pill every day. They thought the medication was for both of them. Nobody had thought to explain how the pills worked, or at least, that only the wife was supposed to take them, showing that the soft aspects of technology design – instructions, in this case – really do matter.

As one recipient of foreign aid concisely noted: "The donors give us what they have, not what we need" – a lack of Human-tech vision that has led governments and organizations to invest about a trillion US dollars of largely unsuccessful aid in the Third World over the last fifty years.[18]

Human development can't be viewed as a "plug and play" activity where technology is airlifted from one part of the world to the other. Context matters. The key to a successful technology *isn't* the technology itself, but rather its affinity with its users. Nor is what's best for the rich necessarily what's best for the poor. Less developed countries have different needs, and simply don't have the social structures in place to take advantage of complex technology from developed countries. The focus needs to be on a different kind of technology, one that has affinity with the human nature *of the people who will be using it*. In many countries, this will mean technology that is simple, cheap, flexible and adapted to unique local needs. At the same time, however, innovations in the design of "soft" technological systems can be introduced at a rate that allows underdeveloped countries to pick up the knowledge they will need to take advantage of more complex technologies as, and when, they can use them. Mind you, none of this is new. People like Fritz Schumacher who have studied underdeveloped countries up close, using an approach that is Human-tech in all but name, have already come to the same conclusion.[19]

The *media* also have a critical role to play. Plenty of news coverage is devoted to technology stories, but the human factor is frequently missing – a sure sign of the traditional Mechanistic perspective. On the rare occasions when the relationship between people and technology is addressed, machines are too often portrayed as "intelligent" whereas people are portrayed as incompetent – the kind of distorted coverage that feeds the insecurities many people already feel about being technological "dummies," and bolsters obsolete kinds of thinking. That's because the idea of making people and their needs the central focus of technological development is so foreign to most people. The prevailing view of the role of technology in society treats people as stupid and sees "design for dummies" as the solution to our frustrations and difficulties. It doesn't treat people with the dignity that they deserve.

The media do occasionally take the opportunity to play a powerful role in moving us into the Human-tech age – a meta-technical age – by educating the public on the enormous impact that such thinking can have on our quality of life. The Canadian television program *W-5* about lethal vincristine errors is a good example. In a scant twenty-five minutes (including commercials), the way preventable medical error can affect each of us and our loved ones was laid out in all of its unfor-gettable human drama. Anyone who watched that program and saw what eight-year-old Ryan Bishop, four-year-old Courtney Braund, seven-year-old Kristine Walker and their families went through can't help but come away thinking that developing Human-tech thinking in health care should be at the very top of politicians' agendas, because of the tremendous benefit it can have for society.

But media coverage of this type is still the exception, not the norm. Let me give you just two examples. In November 2001, I was invited by the Ontario Hospital Association Convention to give a presentation on how to design medical devices that are easier for health-care professionals to use. While waiting to give my talk, I found a stand with a stack of copies of a special market-ing supplement that had appeared in the previous week's *Globe and Mail*, one of Canada's largest daily newspapers. Apparently, the OHA had paid for the supplement to advertise its annual convention, and they were now giving away copies of this sup-plement to the convention participants. Since the theme of the convention was "Touching Technology," there were several arti-cles about technology in health care, and I read them all with interest to see if I could find some relevant background materi-al for my own presentation. I pulled the following five sentences from two separate articles in the same eight-page supplement:[20]

- "We are using the new technology to allow hospital staff more time to focus on and provide that vital human touch."
- "Today there is so much high technology in almost any operation that there is no way even the best of nurses can understand all of it."
- "Technology places incredible burdens on every nursing departments [sic]."
- "The burden is certain to grow and perhaps never lessen."
- "Technology is the hope for the future of health care."

Apparently, the editors of the supplement didn't notice the astounding contradictions contained in these five statements – something that would not have surprised the brilliant Arthur Koestler, who so well understood the psychological roadblocks that impede our vision of reality.

Lest you think that such muddled thinking is an ailment that's specific to health care, let me give you a second example from an entirely different realm – the public debate in early 2001 surrounding U.S. plans to create missile defence technology. Regardless of which side you're on, you should have some semblance of a coherent argument to back up your position, especially if you're the Secretary of State. Right? Apparently not. Colin Powell was quoted as saying: "We are not going to be knocked off the track of moving in this direction as long as the technology points us in that direction."[21]

I don't know if this is an accurate statement of Powell's views, but let's just assume that it is. It's a remarkable position to take. What he said, in essence, was "If we can build a missile defence technology, then we will." Take a second to think about that. Because the same argument could be used to justify *any* technology that can be built – like a concentration camp with efficient gas chambers, as the Canadian Nobel laureate in chemistry, John Polanyi, once pointed out to me. Technology in fact points in almost any direction you can think of. It can let you do all kinds of things, but if those things don't take the human factor into account, why do them? Many technologically feasible things are at best stupid and at worst apocalyptic. Of course, the reporter who wrote this story isn't responsible for Powell's views and is required to report them, not doctor them, but it's unfortunate that the op-ed editors of newspapers so rarely notice – or comment on – the nonsensical statements that arise from such a one-eyed Mechanistic approach.

Finally, *universities* can also contribute by taking a critical look at technical education. Since I'm an engineering professor, I'll concentrate on engineering education. Most designers focus on technology because engineering curricula are largely still based on a Mechanistic world view. Students learn about thermodynamics, materials science, calculus, linear algebra, chemistry, physics, electrical circuits. All these topics are important; our planes wouldn't fly and our bridges wouldn't stand without them. But, as I've tried to show, these skills aren't enough any more. The Human-tech Revolution doesn't do away with the need for technical competence; designing to fit the physical world *is* critical. It just asks that we go beyond technical excellence, that we look at the interaction between people and technology. And, surprising as it may be, most engineering students are *never* taught to do that. They graduate without having taken a single course in designing *for people*. They join the workforce without being skilled in Human-tech thinking. They live in a Mechanistic world where they literally can't see incompatibilities between people and technology. The best and the brightest may overcome these educational deficiencies and learn the value of Human-tech thinking from practical experience, but most don't. And *they* wind up designing very impressive technical widgets that most people use with difficulty, if at all.

The impact of educational change is slow, but all engineers need to know that it's possible to design technology that has an affinity with human nature. They need to know that there are systematic methods for achieving that goal, and that those methods have been proven to make a difference. They need to know the tremendous negative consequences of focusing on technical details alone. Mind you, that doesn't mean that all engineers should become Human-tech experts. That's not realistic, nor desirable for that matter. But they do need to know that attention to the interaction between people and technology is a crucial part of good design and that there are experts who specialize in that area.

It doesn't take much, actually. For example, almost every engineering undergraduate is required to take one course in engineering economic analysis. That doesn't make these students expert accountants or economists; specialists will always be needed. But taking the course exposes students to the relevance of economic factors to project management, and engineers are required to take it because, no matter what the industry, every engineering project has a budget: you simply can't avoid economic considerations. The very same argument applies to the human factor: every engineering project involves interaction between technology and people somewhere along the way. Yet there's currently no requirement that I know of for all engineers to be exposed to these considerations.

I also believe that, in the long run, making this kind of change to engineering education would result in tremendous – possibly surprising – benefits. Right now, engineering tends to attract people who are born Mechanistic thinkers – technologically clever, but somewhat narrow in their interests. Most of them are men. A lot of students who are just as good at the technical details but also have broad interests don't go into engineering, or if they do, don't stick with it, because they think it will be boring to spend four years taking only math, science and technical courses. These students are at least as bright as the ones who do currently go into and stay in engineering. They too have excellent high school grades, especially in maths and sciences. But they have wide interests. They can write a proper English sentence.

They can speak clearly and convincingly. They're heavily involved in extracurricular activities. They have good people skills. They read newspapers. They're leaders. They want to learn about history, psychology, politics, sociology and other courses that don't currently have a place in the rigid and overly prescriptive engineering curricula. These students – ones that we're currently losing to other disciplines – would make outstanding engineers and precious leaders in society. They're born Human-tech thinkers. Many of them are women.

By putting more emphasis on Human-tech thinking we would be graduating better engineers. Technical skills are essential, sure, but many of society's problems demand a broader view, and many of the students we graduate are ill prepared for the challenges that await them. They wind up perpetuating rather than solving our societal problems. And the students who bypass engineering and become leaders often don't have the knowledge or the skills they need to tackle society's important problems – many of which are technological in nature, given the degree to which technology is becoming central to so many critical sectors, and not just at the physical level, but at the level of organizations and political systems too.[22] Most people in leadership positions in government, for instance, have no technical background at all; many are graduates of law, politics or business. (Can you name one prominent politician who has an engineering degree? Answer: Jimmy Carter.) Today's leaders find their way into positions of power and influence because they have a good sense of history, context, people and organizations – knowledge that is absolutely essential in making public policy decisions. But it's clear that many of society's problems today also require a knowledge of technology. Do you think that politicians without any technical knowledge are in the best position to make life-critical decisions about the safety of our water, for instance, let alone nuclear or environmental decisions? Society needs a new breed of leader – one that is as comfortable with differential equations and computers as with human psychology and politics.

The history of humankind is filled with impressive conceptual revolutions, such as the Copernican revolution in astronomy, the Darwinian revolution in biology, and the Einsteinian revolution in physics. Whether the Human-tech Revolution will ever have a similar degree of influence, only the future can tell. As the American former baseball great and king of one-liners, Yogi Berra, once said, "Prediction is difficult, especially of the future." But as the management expert and systems thinker Stafford Beer observed, "The future is something we use our freedom to determine."[23] I think we should use our freedom *now*, to institute Human-tech reform wherever it's needed.

The time has come to concentrate on making the world a better place, and to that end a Human-tech Revolution could have a profound impact on modern times. Technology that has a close affinity with human nature is capable of creating tremendous social changes, on a global scale. It could radically improve our lives and those of our children and grandchildren.

Anything else will just prolong the period of transitional instability we're mired in, leading to more airplanes crashing down to earth; more nuclear power plant meltdowns; more space shuttle explosions; more public health epidemics that quickly race across international borders; more oil tanker accidents destroying our natural environment and more four-year-old children killed by preventable medical errors.

Notes

Preface

I   In North America, the term "human factors engineering" is sometimes used to refer to psychological work, whereas the term "ergonomics" is sometimes used to refer to physical work, but more often than not, the two are treated as synonyms. In Europe and elsewhere, "ergonomics" is the preferred term and is used to refer to both the physical and the psychological.

Chapter One

I   This account of the Chernobyl accident is based on the following sources: "Chernobyl" (World Nuclear Association, 2001), available at www.world-nuclear.org/info/chernobyl/inf07.htm, 14 December 2001; J. Reason, *Human Error* (Cambridge, England: Cambridge University Press, 1990); "Chernobyl fifteen years on: Thoughts and lessons from the Chernobyl accident," Nuclear Energy Association *USA Health Physics Society Newsletter* (April 2000), (www.nea.fr/html/rp/chernobyl/allchernobyl.html, 14 December 2001); P. Gould, *Fire in the Rain: The Democratic Consequences of Chernobyl* (Cambridge, England: Polity Press, 1990); P. P. Read, *Ablaze: The Story of Chernobyl* (London: Secker and Warburg, 1993).

2   A number of other factors also contributed to the accident. See
    Reason, *Human Error.*

3   "Chernobyl," World Nuclear Association; Reason, *Human Error;*
    "Chernobyl, Fifteen Years On," Nuclear Energy Association.

4   U. Franklin, *The Real World of Technology,* revised edition
    (Toronto: House of Anansi Press, 1999).

5   R. W. Lucky, "Design for people – NOT," *IEEE Spectrum* 36(7):
    20–21 (1999).

6   A. Lightman, *The World Is Too Much with Me: Finding Private
    Space in the Wired World* (Toronto: Hart House, University of
    Toronto, 2002); J. Gleick, *Faster: The Acceleration of Just About
    Everything* (New York: Pantheon, 2002).

7   L. T. Kohn, J. M. Corrigan, and M. S. Donaldson, *To Err Is
    Human: Building a Safer Health System* (Washington, DC:
    National Academy Press, 1999).

8   T. A. Brennan et al., "Incidence of Adverse Events and
    Negligence in Hospitalized Patients: Results of the Harvard
    Medical Practice Study I," *New England Journal of Medicine* 324
    (1991): 370–76. L. L. Leape et al., "The Nature of Adverse Events
    in Hospitalized Patients: Results of the Harvard Medical Practice
    Study II" *New England Journal of Medicine* 324 (1991): 377–84.

9   R. M. Wilson et al., "The Quality in Australian Health Care
    Study," *Medical Journal of Australia* 163 (1995): 45–71; C. Vincent
    et al., "Adverse Events in British Hospitals: Preliminary
    Retrospective Record Review," *British Medical Journal* 322
    (2001): 517–19; P. Davis et al., "Adverse Events in New Zealand
    Public Hospitals I: Occurrence and Impact," *New Zealand
    Medical Journal* 115 (2001) 271–80.

10 T. Brennan, "The Institute of Medicine Report on Medical Errors
    – Could It Do Harm?" *New England Journal of Medicine* 342
    (2002), 1123–25; C. Newhall, letter to the editor, *New England
    Journal of Medicine* 343 (2000), 664; W. C. Richardson, D. M.
    Berwick and J. C. Bisgard, letter to the editor, *New England
    Journal of Medicine,* 343 (2002), 663–64; C. J. McDonald, M.
    Weiner, and S. L. Hui, "Deaths Due to Medical Errors are

Exaggerated in Institute of Medicine Report," *Journal of the American Medical Association* 284 (2000): 93–95; L. L. Leape, "Institute of Medicine Medical Error Figures Are Not Exaggerated," *Journal of the American Medical Association* 284 (2000), 95–97.

11 Kohn et al., 3.

12 D. B. Guralnik, *Webster's New World Dictionary of the American Language,* Second College Edition (Cleveland: William Collins and World Publishing, 1976), 1460.

13 Neil Postman, *Technopoly: The Surrender of Culture to Technology* (New York: Vintage, 1992).

14 D. M. Gaba and S. K. Howard, "Fatigue Among Clinicians and the Safety of Patients," *New England Journal of Medicine* 347 (2002): 1249–55.

15 "Pilot Fatigue, Error Probably Causes of '99 Little Rock Crash," CNN, 2001 (www.cnn.com/2001/US/10/23/little.rock.crash/index.html, 26 April 2002).

16 Gaba and Howard.

17 R. Steinbrook, "The Debate over Residents' Work Hours," *New England Journal of Medicine* 347 (2002): 1296–1302.

18 L. O'Brien-Pallas et al., "The Economic Impact of Nurse Staffing Decisions: Time to Turn Down Another Road?" *Hospital Quarterly* 4(3): 42–50 (2001).

19 C. Kovner and P. J. Gergen, "Nursing Staff Levels and Adverse Events Following Surgery in U.S. Hospitals," *Image: Journal of Nursing Scholarship* 30 (1998), 315–21; L. Aiken et al., "Hospital Nurse Staffing and Patient Mortality, Nurse Burnout, and Job Dissatisfaction," *Journal of the American Medical Association* 288 (2002): 1987–93; J. Needleman et al., "Nurse-Staffing Levels and the Quality of Care in Hospitals," *New England Journal of Medicine* 346 (2002): 1715–22.

20 National Transportation Safety Board, "Accidents, Fatalities, and Rates, 1982 through 2000, for U.S. Air Carriers Operating Under 14 CFR 121, Scheduled Service (Airlines)" (www.ntsb.gov/aviation/Table6.htm, 12 December 2001).

21   T. S. Perry, "In Search of the Future of Air Traffic Control," *IEEE Spectrum* 34(8): 19–35 (1997).

22   P. A. Hancock and S. G. Hart, "Defeating Terrorism: What Can Human Factors/Ergonomics Offer?" *Ergonomics in Design* 10(1): 6–15 (2002).

23   N. Moray, "Technosophy and Humane Factors," *Ergonomics in Design* 1(4): 33–39 (1993); N. Moray, "Ergonomics and the Global Problems of the 21st Century," keynote address presented at the 12th Triennial Congress of the International Ergonomics Association, Toronto, August 1994; R. Nickerson, *Looking Ahead: Human Factors Challenges in a Changing World* (Hillsdale, N.J.: Erlbaum, 1992); R. S. Nickerson and N. P. Moray, "Environmental change," in R. S. Nickerson, ed., *Emerging Needs and Opportunities for Human Factors Research* (Washington, D.C.: National Academy Press, 1995): 158–76; M. Strong, *Where On Earth Are We Going?* (Toronto: Knopf Canada, 2000); D. Peterson, "Safe Drinking Water for Rural Canada," in *Proceedings of the Conference on Safe Drinking Water Production in Rural Areas: Issues, Challenges and Future Directions* (Saskatoon, Sask..: Safe Drinking Water Foundation, 2002) (www.safewater.org/conferences/proceedings/peterson.htm, November 29, 2002) World Resource Institute Facts and Figures: Environmental Data Tables (www.wri.org/facts/data-tables-forests.html, March 7, 2003.)

24   Smallpox was eradicated in 1979, but in its last hundred years it killed about one billion people, outranking the Black Plague in terms of human carnage. It's widely considered to be the worst human disease of all because its effects are so lethal and its spread so rapid and so wide. Many of us were inoculated against smallpox as children – creating a scar on our upper arm – but the vaccine's effects start to wear off after about five years, so the vast majority of the world's population is still susceptible to the virus's deadly and ghastly effects. And because our planet is now more connected than ever by integrated transportation systems, the disease could spread like wildfire; it took twenty years for the AIDS virus to infect 50 million people, but the strategic use of

smallpox as a biological weapon could infect the same number of people in ten to twenty weeks.

To make matters worse, the remedy itself is not only time-sensitive but also potentially lethal. The smallpox vaccine has to be given within four or five days of exposure to the virus to have any effect. Even then it can cause brain disease and kill about one out of every million people, so it shouldn't be given to 20 per cent of the population, particularly those with weak immune systems (e.g., chemotherapy patients and people suffering from HIV, inflammatory diseases or eczema). Therefore, the need for safe and effective mass inoculation of the population in the face of a large-scale bioterrorist attack using smallpox – an unprecedented event in the history of human civilization – would create non-trivial logistical obstacles and introduce a unique and significant potential for lethal medical error. See R. Preston, *The Demon in the Freezer: A True Story* (New York: Random House, 2002).

Chapter Two

1   C. P. Snow, *The Two Cultures* (1959; Cambridge, England: Cambridge University Press, 1998): 3, 11, 16, 50.

2   E. Hutchins, *Cognition in the Wild.* (Cambridge, Mass.: MIT Press, 1995).

3   D. A. Norman, *The Psychology of Everyday Things* (New York: Basic Books, 1988).

4   Ibid.

5   M. Gooderham, "Baffled Consumers Plead for Simplicity," *Globe and Mail,* 8 December 1998, C1.

6   J. Hopkins, "When the Devil Is in the Design," *USA Today,* 31 December 2001 (www.usatoday.com/money/retail/2001–12–31-design.htm, December 31, 2001).

7   A. Robinson, "BMW 745I: The Ultimate Interfacing Machine," *Car and Driver* (June 2002): 71–75.

8   A. Bornhop, "BMW 745I: iDrive? No, you drive, while I fiddle with the controller," *Road and Track* (June 2002): 74–79.

9   P. Beynon-Davies, "Information Systems 'Failure': The Case of the London Ambulance Service's Computer Aided Despatch Project," *European Journal of Information Systems* 4(1995):171–84; D. Page, P. Williams, and D. Boyd, "Report of the Public Inquiry into the London Ambulance Service" (London: South West Thames Regional Health Authority, 1993).

10   Ibid., 14.

11   Ibid., 5.

12   Beynon-Davies, "Information Systems Failure."

13   Norman, *Psychology of Everyday Things.*

14   Robert Wright, *Nonzero: The Logic of Human Destiny* (New York: Vintage Books, 2000).

15   Ibid., 27.

16   Ibid., 122.

17   J. J. Gibson, *The Ecological Approach to Visual Perception* (Boston: Houghton-Mifflin, 1979).

18   The levels in table 1 are not unique. For example, the "Physical" level could be subdivided into Anatomy and Physiology, and the "Political" level could be subdivided into Public Opinion, Government and Regulatory Associations. Also, the exact number of levels that are useful may differ across sectors (e.g., health care vs. nuclear power). See N. Moray, "Error Reduction as a Systems Problem," in M. S. Bogner, ed., *Human Error in Medicine* (Hillsdale, N.J.: Erlbaum, 1994), 67–91; and J. Rasmussen, "Risk Management in a Dynamic Society," *Safety Science* 22 (1997): 183–213.

19   National Research Council, *Musculoskeletal Disorders and the Workplace: Low Back and Upper Extremities* (Washington, D.C.: National Academy Press, 2001).

20   C. D. Wickens, *Engineering Psychology and Human Performance* (New York: Harper-Collins, 1992).

21   L. Segal, "Designing Team Workstations: The Choreography of Teamwork," in P. Hancock, J. Flach, J. Caird, and K. Vicente, eds., *Local Applications of the Ecological Approach to Human-Machine Systems* (Hillsdale, N.J.: Erlbaum, 1995), 392–415.

22 I. L. Janis, *Victims of Groupthink* (Boston: Houghton-Mifflin, 1972).

23 P. Ayton, and H. Arkes, "Call It Quits," *New Scientist* 158 (1998): 40–43.

24 G. Hardin, "The Tragedy of the Commons," *Science* 162 (1968): 1243–48.

25 K. J. Vicente, *Cognitive Work Analysis: Toward Safe, Productive, and Healthy Computer-Based Work* (Mahwah, N.J.: Erlbaum, 1999).

Chapter Three

1 Webb Associates, *Anthropometric Source Book: Anthropometry for Designers* 1–3, NASA Reference Publication 1024 (Washington, D.C.: NASA, 1978).

2 K. H. E. Kroemer, H. I. Kroemer, and Ebert K. E. Kroemer, *Engineering Physiology: Physiological Bases of Human Factors/Ergonomics* (Amsterdam: Elsevier, 1986).

3 D. Gaba, "Structural and Organizational Issues in Patient Safety: A Comparison of Health Care to Other High-Hazard Industries," *California Management Review* 43 (2000): 83–102; C. Perrow, *Normal Accidents: Living with High-Risk Technologies* (New York: Basic Books, 1984).

4 Reuters, "Airline Disasters Fall," *Metro Today*, 3 January 2002, 1.

5 Although some of the challenges faced by military pilots are obviously more demanding than those faced by their commercial airline counterparts, the deficiencies found in the military planes reflected the state-of-the-art thinking in aviation as a whole. Indeed, many of the design defects documented in military aircraft could also be found in commercial aircraft designs of the day.

6 P. M. Fitts, *Psychological Research on Equipment Design*, Report No. 19 (Dayton, Ohio: Aero Medical Laboratory, Air Materiel Command, Wright-Patterson Air Force Base, 1947).

7 P. M. Fitts and R. E. Jones, "Analysis of Factors Contributing to 460 'Pilot Error' Experiences in Operating Aircraft Controls," Memorandum Report TSEAA-694-12 (Dayton, Ohio: Aero

Medical Laboratory, Air Materiel Command, Wright-Patterson Air Force Base, 1947); P. M. Fitts and R. E. Jones, "*Psychological Aspects of Instrument Display. I: Analysis of 4270 'Pilot Error' Experiences in Reading and Interpreting Aircraft Instruments,*" Memorandum Report TSEAA-694–12A (Dayton, Ohio: Aero Medical Laboratory, Air Materiel Command, Wright-Patterson Air Force Base, 1947).

8   Fitts and Jones, "Analysis of Factors Contributing to 460 'Pilot Error' Experiences in Operating Aircraft Controls," 18.

9   Ibid., 26.

10   S. N. Roscoe, *The Adolescence of Engineering Psychology* (Santa Monica, Calif.: Human Factors and Ergonomics Society, 1997).

11   A. R. Duchossoir, *The Fender Stratocaster: The Success Story of A Legendary Guitar Born and Made in California* (Milwaukee, Wis.: Hal Leonard Publishing, 1988).

12   Ibid., 37.

13   Christie's Auction House, *A Selection of Eric Clapton's Guitars in Aid of the Crossroads Centre* (New York, 1999).

14   A. R. Duchossoir, *The Fender Telecaster: The Detailed Story of America's Senior Solid Body Electric Guitar* (Milwaukee, Wis.: Hal Leonard Publishing, 1991).

15   Duchossoir, *Fender Stratocaster,* 2.

16   Ibid.

17   Ibid., 5.

18   Ibid.

19   Ibid., 6.

20   Ibid., 18.

21   Ibid.

22   Duchossoir, *Fender Telecaster,* 7.

23   Guitar Player (columnist), "The Hendrix Interviews: One of His First. October '68," *Guitar Heroes* 1(1) (1992): 49.

24   J. R. Guilfoyle, "What Design Has Done for the Toothbrush," *Industrial Design* (November/December 1977: 34–38).

25   G. Smets, "Industrial Design Engineering and the Theory of Direct Perception and Action," *Ecological Psychology* 7 (1995): 329–74.

Chapter Four

1   D. A. Norman, *The Psychology of Everyday Things* (New York: Basic Books, 1998); C. D. Wickens, S. E. Gordon, and Y. Liu, *An Introduction to Human Factors Engineering* (Reading, Mass.: Addison-Wesley, 1997).

    This discipline also goes by several other names that have different connotations to insiders, but that for our purposes, are roughly equivalent: cognitive ergonomics, engineering psychology, and cognitive engineering.

2   www.swindonweb.com/life/lifemagi0.htm.

3   Alphonse Chapanis and L. Lindenbaum, "A Reaction Time Study of Four Control-Display Linkages," *Human Factors* 1(4)(1950): 1–7.

4   E. Bergman and R. Haitani, "Designing the PalmPilot: A Conversation with Rob Haitani," in E. Bergman, ed., *Information Appliances and Beyond* (San Francisco, Calif.: Morgan Kaufmann, 2000, 81–102).

5   Ibid., 87.

6   Ibid., 83.

7   Ibid., 95.

8   www.palm.com.

9   W. S. Cleveland, *The Elements of Graphing Data* (Monterey, Calif.: Wadsworth, 1985).

10 P. Underhill, *Why We Buy: The Science of Shopping* (New York: Simon and Schuster, 1999).

11 Ibid., 17.

12 Ibid., 43.

13 Ibid., 37.

14 Ibid., 82.

Chapter Five

1   These examples were already presented in my "Human Factors and Global Problems: A Systems Perspective" (*Systems Engineering* 1 (1998): 57–69.

2   D. Kuk, J. Cowley, and F. A. Beserve, "Human Factors: Two Different Directions in Energy Conservation," unpublished

manuscript (Toronto: Department of Industrial Engineering, University of Toronto, 1994).

3   C. Long, D. Ramsahai, Z. Bhujwalla, and J. Wright, "Redesigning Photocopiers to Reduce Paper Wastage," unpublished manuscript (Toronto: Department of Industrial Engineering, University of Toronto, 1994).

4   J. L. Seminara, W. R. Gonzalez, and S. O. Parsons, *Human Factors Review of Nuclear Plant Control Room Design*, Electric Power Research Institute, Palo Alto, Calif. (1977): NP-309.

5   John G. Kemeny, *Report of the President's Commission on the Accident at Three Mile Island: The Need for Change: The Legacy of TMI*, Government Printing Office, Washington, D.C. (1979): 112.

6   Seminara et al., 1977.

7   GPU Nuclear Corporation, Three Mile Island, Communications Department.

8   J. M. Broughton, P. Kuan, and D. A. Petti, "A Scenario of the Three Mile Island Unit 2 Accident," *Nuclear Technology* 87 (1989): 34–53.

9   E. Rubinstein, "The accident that shouldn't have happened," *IEEE Spectrum* 16 (1979): 33–42.

10 L. Beltracchi, "A Direct Manipulation Interface for Water-based Rankine Cycle Heat Engines," *IEEE Transactions on Systems, Man, and Cybernetics* SMC-17 (1987), 478–87.

11 K. J. Vicente et al., "Evaluation of a Rankine Cycle Display for Nuclear Power Plant Monitoring and Diagnosis" *Human Factors* 38 (1996): 506–21.

12 International Atomic Energy Association, "Electricity, Nuclear Power and the Global Environment – Fact Sheet." (www.iaea.org/ worldatom/Periodicals/Factsheets/English/, 23 December 2002).

13 This description of the problems with airport security is based on the following references: P. A. Hancock and S. G. Hart, "Defeating Terrorism: What Can Human Factors/Ergonomics Offer?" *Ergonomics in Design* 10(1): 6–16 (2002); D. Harris, "How to *Really* Improve Airport Security," *Ergonomics in Design* 10(1): 17–22 (2002); G. Kaempf, D. Klinger, and S. Wolf, *Development of*

Decision-Centered Interventions for Airport Security Checkpoints, final technical report (Yellow Springs, Ohio: Klein Associates Inc., 1994); G. Kaempf, D. Klinger, and S. Wolf, *Performance Measurement for Airport Security Personnel,* final technical report (Yellow Springs, Ohio: Klein Associates Inc., 1996).

14 Harris, "Airport Security," 20.

15 D. R. Davies and R. Parasuraman, *The Psychology of Vigilance* (London: Academic, 1982).

16 D. Ose, *Briefing Memorandum for November 27, 2001 Hearing, "What Regulations Are Needed to Ensure Air Security?"* (Washington, D.C.: Committee on Government Reform, House of Representatives, Congress of the United States, 2001).

17 Kaempf et al., "Performance Measurement," 12–13.

18 Hancock and Hart, "Defeating Terrorism."

19 The tragic details of this case were obtained by making a Freedom of Information Act request of the Tallahassee State Attorney. See W. N. Meggs and W. Hicks, *Investigation – Death of Danielle Rone McCray,* SID 2000–03 (Tallahassee, Fla.: Office of State Attorney, Second Judicial Circuit, Special Investigation Division, 2000).

20 K. J. Vicente, K. Kada-Bekhaled, G. Hillel, A. Cassano, and B. A. Orser, "Programming Errors Contribute to Death from Patient-Controlled Analgesia: Case Report and Estimate of Probability," *Canadian Journal of Anesthesia* 50(4)(2003): 328–32.

21 Institute for Safe Medication Practices, "Evidence Builds: Lack of Focus on Human Factors Allows Error-Prone Devices," *ISMP Medication Safety Alert* 4(15)(1999); M. B. Weinger et al., "Incorporating Human Factors into the Design of Medical Devices," *Journal of the American Medical Association* 280 (1998): 1484; Emergency Care Research Institute. "Abbott PCA Plus II Patient-Controlled Analgesia Pumps Prone to Misprogramming Resulting in Narcotic Overinfusions," *Health Devices* 26 (1997): 389–91.

22 Emergency Care Research Institute, "Abbott PCA Plus II," 390.

23 Emergency Care Research Institute, "Patient-Controlled Analgesic Infusion Pumps," *Health Devices* 30 (2001): 172.

24   C. H. McLeskey, "Abbott Addresses Medication Errors through Advanced PCA Technology," *APSF Newsletter* 15 (2000): 36–37. (www.gasnet.org//societies/apsf/newsletter/2000/fall/07OpinionResponse.htm, 22 August 2001); Abbott Laboratories, "LifeCare® PCA Plus II Infuser" (www.abbotthosp.com/PROD/pain/pcaplus.html, 23 August 2001).

25   Emergency Care Research Institute, "Medication Safety: PCA Pump Programming Errors Continue to Cause Fatal Overinfusions," *Health Devices* 31 (2002): 342–47.

26   Vicente et al., "Programming Errors Contribute to Death."

27   Institute for Safe Medication Practices, "Safety briefs," *ISMP Medication Safety Alert* 1(9)(1996): 1.

28   L. Lin et al., "Applying Human Factors to the Design of Medical Equipment: Patient-Controlled Analgesia," *Journal of Clinical Monitoring and Computing* 14 (1998): 253–63. (www.kluweronline.com/issn/1387–1307, 25 February 2002.)

29   L. Lin, K. J. Vicente, and D. J. Doyle, "Patient Safety, Potential Adverse Drug Events, and Medical Device Design: A Human Factors Engineering Approach," *Journal of Biomedical Informatics* 34 (2001): 274–84 (www.idealibrary.com/links/doi/10.1006/jbin.2001.1028/pdf, 28 March 2002).

30   D. Mayer, "Dear Clinician," letter dated 29 October 1997 (Hospital Products Division of Abbott Laboratories, 1997).

31   P. St. John, "Drug Pump's Deadly Trail," *Tallahassee Democrat* (28 May 2001) A1.

32   C. H. McLeskey, "Patient-Controlled Analgesia," *Canadian Medical Association Journal* 164 (2001): 620–21; Vicente et al., "Programming Errors Contribute to Death."

33   Institute for Safe Medication Practices, "Safety Briefs," ISMP Medication Safety Alert, 1996: 1(12); C. H. McCleskey, "Abbott Addresses Medication Errors" (2000) and "Patient-Controlled Analgesia" (2001).

34   I've been told that the manufacturer has since made a tremendous turnaround, hiring a human factors engineering program manager at corporate headquarters, setting up a human factors

council with representatives from each of the corporation's divisions, creating a human factors design process for creating all future medical devices, and putting on training courses throughout the company to convince designers of the importance of creating a tight fit between people and technology.

35   L. L. Leape, "Error in Medicine," *Journal of the American Medical Association* 272 (1994): 1851–57.

36   The remarkable and inspiring story of how Jane became a practising physician, after becoming blind from juvenile diabetes just a few weeks before her medical exams, is told in: Jane Poulson, *The Doctor Will Not See You Now: The Autobiography of a Blind Physician* (Ottawa: Novalis, 2002).

37   See Meggs and Hicks, "Death of Danielle Rone McCray," 3ff.

38   Leape, 1857.

39   J. B. Cooper, R. S. Newbower, C. D. Long., and B. McPeek, "Preventable Anesthesia Mishaps: A Study of Human Factors," *Anesthesiology* 49 (1978): 399–406.

40   Ibid., 404.

41   Ibid., 402.

42   Arthur Koestler, *The Sleepwalkers: A History of Man's Changing Vision of the Universe* (London: Penguin, 1959).

Chapter Six

1   National Transportation Safety Board, *Aircraft accident report: Eastern Air Lines, Inc. L-1011, N310EA, Miami, Florida, December 29, 1972,* NTSB-AAR-73-14 (Washington, D.C.: 1973).

2   The autopilot becomes disengaged as soon as a pilot grabs the control yoke to fly the plane manually. In this case, one of the crew members likely brushed up against the control yoke without noticing.

3   R. L. Helmreich, and H. C. Foushee, "Why Crew Resource Management? Empirical and Theoretical Bases of Human Factors Training in Aviation," in E. L. Weiner, B. G. Kanki, and R. L. Helmreich, eds., *Cockpit Resource Management* (San Diego, Calif.: Academic Press, 1993) 3–45.

4 National Transportation Safety Board, *Aircraft Accident Report: United Airlines, Inc. McDonnell-Douglas DC-8-61, N8082U, Portland, Oregon, December 28, 1978,* NTSB-AAR-79-7 (Washington, DC: 1979).

5 H. C. Foushee, "Dyads and Triads at 35,000 Feet: Factors Affecting Group Process and Crew Performance," *American Psychologist* 39 (1984): 885–93.

6 Ibid.

7 H. W. Orlady, "Airline Pilot Training Today and Tomorrow," in E. L. Weiner, B. G. Kanki, and R. L. Helmreich, eds., *Cockpit Resource Management,* 447–77.

8 Weiner, Kanki, and Helmreich, *Cockpit Resource Management.*

9 R. E. Butler, "LOFT: Full-Mission Simulation as Crew Resource Management Training," in Weiner, Kanki, and Helmreich, *Cockpit Resource Management,* 231–59.

10 L. Segal, "Designing Team Workstations: The Choreography of Teamwork, in P. Hancock et al., eds., *Local Applications of the Ecological Approach to Human-Machine Systems* (Hillsdale, N.J.: Erlbaum, 1995), 392–415.

11 R. E. Byrnes, and R. Black, "Developing and Implementing CRM Programs: The Delta Experience," in Weiner, Kanki, and Helmreich, *Cockpit Resource Management,* 421–43.

12 R. L. Helmreich, E. L. Weiner, and B. G. Kanki, "The Future of Crew Resource Management in the Cockpit and Elsewhere, in Weiner, Kanki, and Helmreich, *Cockpit Resource Management,* 479–501.

13 Ibid.

14 R. L. Helmreich, and A. C. Merritt, *Culture at Work in Aviation and Medicine: National, Organizational and Professional Influences* (Aldershot, England: Ashgate, 1998).

15 R. L. Helmreich, and H.-G. Schaefer, "Team Performance in the Operating Room," in M. S. Bogner, ed., *Human Error in Medicine* (Hillsdale, N.J.: Erlbaum, 1994), 225–53.

16 L. Lingard et al., "Team Communication in the Operating Room: Talk Patterns, Sites of Tension, and Implications for Novices," *Academic Medicine* 77 (2002), 232–37.

17   Helmreich and Merritt, *Culture at Work in Aviation*, 169–70.

18   S. K. Howard, D. M. Gaba, K. J. Fish, G. Yang, and F. H. Sarnquist, "Anesthesia Crisis Resource Management Training: Teaching Anesthesiologists to Handle Critical Incidents," *Aviation, Space & Environmental Medicine* 63 (1992): 763–70; D. M. Gaba et al., 'Simulation-Based Training in Anesthesia Crisis Resource Management (ACRM): A Decade of Experience," *Simulation & Gaming* 32 (2001): 175–93.

19   D. M. Gaba, and A. DeAnda, "A Comprehensive Anesthesia Simulation Environment: Re-Creating the Operating Room for Research and Training," *Anesthesiology* 69, 387–94. For a description of the latest version, see: anesthesia.stanford.edu/ VASimulator/sim.htm, 25 March 2002.

20   D. M. Gaba, K. J. Fish, and S. K. Howard, *Crisis Management in Anesthesiology* (New York: Churchill Livingstone, 1994).

21   Gaba et al. "Simulation-based Training in Anesthesia" (2001).

22   Helmreich, Weiner, and Kanki, "Future of Crew Resource Management."

23   N. Johnston, "CRM: Cross-cultural Perspectives," in Weiner, Kanki, and Helmreich, *Cockpit Resource Management.*

24   H. Yamamori, and T. Mito, "Keeping CRM Is Keeping the Flight Safe," in Weiner, Kanki, and Helmreich, *Cockpit Resource Management,* 399–420.

Chapter Seven

1   An increasing number of hospitals have adopted such procedures, after the widely publicized Willie King case – the Florida man who had the wrong leg amputated. For example, one procedure requires that doctors or nurses circle or label the correct limb ahead of time with a marker so that there is a visible indication and reminder of which limb to operate on at the time of surgery.

2   L. T. Kohn, J. M., Corrigan, and M. S. Donaldson, *To Err Is Human: Building a Safer Health System* (Washington, D.C.: National Academy Press, 1999).

3   I will use the term "management" to refer to the people who

supervise or have authority over front-line workers or other managers in an organization (thereby including both middle and upper managers), or to the activities performed by such individuals.

4   This description is based on: D. Vaughan, *The Challenger Launch Decision: Risky Technology, Culture, and Deviance at NASA* (Chicago: University of Chicago Press, 1988).

5   Richard Feynman, *"What do you care what other people think?' Further Adventures of a Curious Character"* (New York: Norton, 1988).

6   J. Rasmussen, "Risk Management in a Dynamic Society: A Modelling Problem," *Safety Science* 27 (1997): 183–213; D. R. O'Connor, *Part One – Report of the Walkerton Inquiry: The Events of May 2000 and Related Issues* (Toronto: Ontario Ministry of the Attorney General, 2002), (www.walkertoninquiry.com, 18 January 2002); R. D. Lang, *Report of the Commission of Inquiry into Matters Relating to the Safety of the Public Drinking Water System in the City of North Battleford, Saskatchewan* (Regina, Sask.: The Queen's Printer, 2002), (www.northbattlefordwaterinquiry.ca, 27 May 2002); J. R. Saul, *Voltaire's Bastards: The Dictatorship of Reason in the West* (Toronto: Penguin, 1992).

7   This account was previously published, in a slightly different form, in K. J. Vicente, "Crazy Clocks: Counterintuitive Consequences of 'Intelligent' Automation," *IEEE Intelligent Systems* 16(6)(2001), 74–76. The original sources are T. S. Perry "Does anybody really know what time it is?" *IEEE Spectrum* 37(10)(2000): 26–28; T. S. Perry "Watching the Clocks," *IEEE Spectrum* 37(12)(2000): 57–60.

8   C. Billings, "Incident Reporting Systems in Medicine and Experience with the Aviation Safety Reporting System," in R. I. Cook, D. D. Woods, and C. Miller, eds., *A Tale of Two Stories: Contrasting Views of Patient Safety* (Chicago: National Patient Safety Foundation at the AMA, 1998), 52–61, E. Billings, "The NASA Aviation Safety Reporting System: Lessons Learned from Voluntary Incident Reporting," in *Proceedings of Enhancing Patient Safety and Reducing Errors in Health Care* (Chicago: National Patient Safety Foundation, 1999), 97–100.

9 Billings, "Lessons Learned from Voluntary Incident Reporting," 59.

10 D. J. Cullen et al., "The Incident Reporting System Does Not Detect Adverse Drug Events: A Problem for Quality Improvement," *Joint Commission Journal of Quality Improvement* 21(10)(1995): 541–52; S. Gardner, and M. Flack, "Designing a Medical Device Surveillance Network," FDA report to Congress 1999 (www.fda.gov/cdrh/postsurv/medsun.html, 23 November 2001); D. C. Classen et al., "Computerized Surveillance of Adverse Drug Events in Hospital Patients," *Journal of the American Medical Association* 266 (1991): 2847–51; A. K. Jha et al., "Identifying Adverse Drug Events: Development of a Computer-Based Monitor and Comparison with Chart Review and Stimulated Voluntary Report," *Journal of American Medical Informatics Association* 5 (1998): 305–14; B. S. Bennett and A. G. Lipman, "Comparative Study of Prospective Surveillance and Voluntary Reporting in Determining the Incidence of Adverse Drug Reactions," *American Journal of Hospital Pharmacy* 34 (1977): 931–36; M. R. Keith, R. A. Bellanger-McCleery, and J. E. Fuchs, Jr., "Multidiscplinary Program for Detecting and Evaluating Adverse Drug Events," *American Journal of Hospital Pharmacy* 46 (1989): 1809–1989.

11 Keith et al., "Detecting and Evaluating Adverse Drug Events."

12 *American Law of Products Liability, 3d, Part 4: Negligence Liability,* Chapter 14: Proof of Breach of Duty (Rochester, N.Y.: The Lawyers Co-operative Publishing, 1987).

13 Ibid., 63.

14 B. Liang, "Error in Medicine: Legal Impediments to U.S. Reform," *Journal of Health Politics, Policy and Law* 24 (1999): 27–58.

15 National Patient Safety Foundation, *Proceedings of Enhancing Patient Safety and Reducing Errors in Health Care* (Chicago: National Patient Safety Foundation, 1999), 65–76.

16 Michael R. Cohen, personal communication, 15 February 2003.

17 Several expert witnesses testified at the trial that the baby might not have died had the pharmacist not made a mistake, but that the combination of the tenfold overdose and the change in route

of administration was certain to lead to death (Michael R. Cohen, personal communication, 13 February 2003).

18   NPSF, *Enhancing Patient Safety,* 73.

19   NSPF, *Enhancing Patient Safety;* Leape (1994); Kohn et al., *To Err Is Human;* D. M. Berwick, "Not again! Preventing Errors Lies in Redesign – Not Exhortation," *British Medical Journal* 322 (2001): 247–48.

20   Of course, there is a grey area between good and bad apples, so it wouldn't always be easy to draw the distinction. But given the health-care sector's current policy of punishing *all* mistakes, this grey area isn't even on the radar. Before we discuss the implementation details of exactly how to distinguish good from bad apples, we have to first recognize that good apples also make mistakes.

21   Berwick, "Not Again!"

22   Ibid., 247.

23   Leape (1994); Kohn et al., *To Err Is Human.*

24   Of course, there are exceptions. As in any profession, it's possible to become callous and insensitive over time.

25   D. Hilfiker, "Facing Our Mistakes," *New England Journal of Medicine* 310 (1984): 118–22.

26   Ibid., 118.

27   Ibid., 119.

28   Ibid.

29   Berwick, "Not Again!" 247.

30   E. M. Tracy, "Evolving Practice and a Culture of Safety," in *Proceedings of Enhancing Patient Safety and Reducing Errors in Health Care* (Chicago: National Patient Safety Foundation, 1999), 38–40.

31   There are a growing number of such programs, as awareness of Human-tech thinking in patient safety is increasing.

32   Tracy, "Evolving Practice."

33   Ibid., 39.

34   Ibid.

35   S. S. Kraman, and G. Hamm, "Risk Management: Extreme

Honesty May Be the Best Policy," *Annals of Internal Medicine* 131 (1999): 963–67.

36   As far as I know, it has not been adopted as policy anywhere else.

Chapter Eight

1   Sun Tzu, *The Illustrated Art of War* (Boston: Shambhala, 1998).

2   Ibid., 84.

3   Thomas Homer-Dixon, "The rise of complex terrorism," *Foreign Policy* (January/February 2002).

4   This description of the history of Hollerith technology and the statistics on its usage are based on: E. Black, *IBM and the Holocaust: The Strategic Alliance between Nazi Germany and America's Most Powerful Corporation* (New York: Crown, 2001).

5   This description of the role that the IBM technology played in the Holocaust is based on the far more detailed account provided by Black in "IBM and the Holocaust."

6   This book is dedicated to him – "Dionisio" was his code name in the days of clandestine activity. This uncle (who is still alive) is from my mother's side of the family, whereas the one that I referred to in chapter 4 – the one who could never remember how to set his VCR clock – comes from my father's side.

7   www.witness.org.

8   S. Hick, E. F. Halpin, and E. Hoskins, *Human Rights and the Internet* (London: MacMillan, 2000).

Chapter Nine

1   *American Law of Products Liability, 3d, Part 4: Negligence Liability,* Chapter 14: Proof of Breach of Duty (Rochester, NY: The Lawyers Co-operative Publishing, 1987).

2   Having said that, public opinion polling numbers can be considered as somewhat comparable in the sense that they're intended to measure human nature at the political level at a particular point in time.

3   Niccolo Machiavelli, *The Prince* (1532; Oxford: Oxford University Press, 1984).

4   W. Shawcross, "The Literary Destruction of Henry Kissinger," *Far Eastern Economic Review* (2 January 1981); Christopher Hitchens, *The Trial of Henry Kissinger* (London: Verso, 2001).

5   Adam Smith, *The Theory of Moral Sentiments* (1759; Indianapolis, Ind.: Liberty Fund, 1984).

6   Taxation is a fourth, but I will not discuss it here.

7   John Ralston Saul, *On Equilibrium* (Toronto: Penguin, 2001), 319.

8   B. Liang, "Error in Medicine: Legal Impediments to U.S. Reform," *Journal of Health Politics, Policy and Law* 24 (1999): 27–58.

9   This analysis of the outbreak is a revised version of Kim J. Vicente and K. Christoffersen (in press), "The Walkerton E. Coli Outbreak: A Test of Rasmussen's Framework for Risk Management in a Dynamic Society," *Theoretical Issues in Ergonomics Science*. The primary sources documenting the events surrounding the outbreak are the 700-page report of the public inquiry and a book written by a Canadian Press journalist: see D. R. O'Connor, *Part One – Report of the Walkerton Inquiry: The Events of May 2000 and related issues* (Toronto: Ontario Ministry of the Attorney General, 2002); D. R. O'Connor, *Part One: A Summary – Report of the Walkerton Inquiry: The Events of May 2000 and related issues* (Toronto: Ontario Ministry of the Attorney General, 2002); C. N. Perkel, *Well of Lies: The Walkerton Water Tragedy* (Toronto: McClelland and Stewart, 2002).

10 Perkel, *Well of Lies.*

11 Ibid., 77.

12 O'Connor, "Summary," 24.

13 Ibid., 7.

14 O'Connor, "Report," 72.

15 Ibid., 62.

16 Ibid., 188.

17 O'Connor, "Summary," 29.

18 O'Connor, "Report," 318.

19 Ibid., 411.

20 O'Connor, "Summary," 35.

21 O'Connor, "Report," 368.

22 O'Connor, "Summary," 33.

23 Smith, *Moral Sentiments,* 233–34.

24 K. J. Vicente and J. Rasmussen, "The Ecology of Human-Machine Systems II: Mediating 'Direct Perception' in Complex Work Domains," *Ecological Psychology* 2 (1990): 207–50; K. J. Vicente and J. Rasmussen, "Ecological Interface Design: Theoretical Foundations," *IEEE Transactions on Systems, Man, and Cybernetics* SMC-22 (1992): 589–606.

25 Adapted from Rasmussen (1997) and reprinted from Vicente (2002). *Quality and Safety in Healthcare, 11,* 302–304. With permission from the BMJ Publishing Group.

26 K. J. Vicente, "Cognitive Engineering Research at Risø from 1962–1979," in E. Salas, ed., *Advances in Human Performance and Cognitive Engineering Research,* vol. 1 (New York: Elsevier, 2001), 1–57.

27 J. Reason, *Human Error* (Cambridge, England: Cambridge University Press, 1990).

28 J. Rasmussen, "Risk Management in a Dynamic Society," *Safety Science* 22 (1997): 183–213.

29 Ibid., 189.

Chapter Ten

1 http://enight.dos.state.fl.us/ 1 February 2001.

2 "Palm Beach Voter Lawsuit Withdrawn As Hearing Convenes." (CNN, 2000), (www.cnn.com/2000/ALLPOLITICS/stories/ 11/09/election.president/index.html, 9 November 2000).

3 R. C. Sinclair, et al., "An Electoral Butterfly Effect," *Nature* 408 (2000): 665–66, (www.nature.com/cgitaf/DynaPage.taf?file=/ nature/journal/v408/n6813/full/408665b0_fs.html, 18 February 2003).

4 "Palm Beach Voter Lawsuit Withdrawn."

5 "Unofficial Florida Tally Completed: Bush's Lead Down to 327" (CNN, 2000), (www.cnn.com/2000/ALLPOLITICS/stories/ 11/10/election.president.03/index.html, 10 November 2000).

6 "Palm Beach Voter Lawsuit Withdrawn."

7 "Palm Beach Commission Approves Computer Voting" (AP, 2001),

(www.cnn.com/2001/US/05/17/electionreform.ap/index.html, 17 May 2001).

8   F. Olsen, "Computer Scientists and Political Scientists Seek to Create a Fiasco-Free Election Day," *The Chronicle of Higher Education* (20 April 2001), chronicle.com; Caltech/MIT Voting Project, *Residual Votes Attributable to Technology: An Assessment of the Reliability of Existing Voting Equipment* (Cambridge, Mass.: MIT, Department of Political Science, 2001).

9   Olsen, "Computer Scientists and Political Scientists."

10 Caltech/MIT Voting Project, "Residual Votes Attributable to Technology," 19, 17.

11 Ibid., 19.

12 D. A. Norman, *The Psychology of Everyday Things* (New York: Basic Books, 1988).

13 I'm indebted to Tim Hurson for this wonderful idea and example. (www.timhurson.com)

14 www.m-w.com/cgi-bin/dictionary.

15 R. Amalberti, "The Paradoxes of Almost Totally Safe Transportation Systems," *Safety Science* 37 (2001): 109–26.

16 In many hospitals, clinical engineering staff are primarily responsible for medical device procurement decisions, but they've tended to evaluate devices only according to strictly technical criteria. One of my students, Andrea Cassano, is working with Dr. Tony Easty, Director of Clinical Engineering at the University of Toronto teaching hospitals, to have Human-tech criteria take centre stage in the procurement process by testing devices with users *before* expensive purchasing decisions are made.

17 J. Stackhouse, *Out of Poverty: And into Something More Comfortable* (Toronto: Random House Canada, 2000).

18 Ibid., 86.

19 E. F. Schumacher, *Small Is Beautiful: Economics As If People Mattered: 25 Years Later . . . With Commentaries* (1973; Point Roberts, Wash.: Hartley and Marks, 1999).

20 "Technology and the Human Touch," *Globe and Mail* (2 November 2001), H1; "Health Care's High-Tech Revolution,"

Globe and Mail (2 November 2001), H5.

21 "U.S. to Ease Missile Defense Fears." (CNN, 10 February 2001), (www.cnn.com/2001/US/02/10/poweel.missile/index.html, 8 August 2001).

22 In 1994, the American Society of Engineering Education put together a report that laid out a vision for engineering education in a changing world. Here's one of the key points made by the panel of experts: "Because engineers now operate in a world where their accomplishments are often more limited by societal considerations than by technical capabilities, they are engaging in a wider range of activities throughout their professional lives. Thus, engineering education must take into account the social, economic, and political contexts of engineering practice." The U.S. National Academy of Engineering has said the same thing. One of its members, Professor D. Allan Bromley, the Dean of Engineering at Yale, convincingly described the need for Human-tech thinking, "I have become increasingly aware that in the average engineering project, the first 10 per cent of the decisions made effectively commit between 80 per cent and 90 per cent of all the resources that subsequently flow into that project. Unfortunately, most engineers are ill-equipped to participate in these important initial decisions because they are not purely technical decisions. Although they have important technical dimensions, they also involve economics, ethics, politics, appreciation of international affairs, and general management considerations. Our current engineering curricula tend to focus on preparing engineers to handle the other 90 per cent, the nut-and-bolt decisions that follow after the first 10 per cent have been made. We need more engineers who can tackle the entire range of decisions." (American Society of Engineering Education, *Engineering Education for a Changing World: Project Report* [Washington, D.C.: 1994], 6; (www.asee.org/publications/reports/greenworld.cfm, 26 April 2002); National Research Council, *Engineering education: Designing an adaptive system* [Washington, D.C.: National Academy Press, 1995], 20.

23 S. Beer, *Designing Freedom* (Concord, Ont.: Anansi, 1973), 91.

Acknowledgments

My greatest debt in writing this book is to Louise Dennys, who also cannot deny some responsibility. She took a risk by believing in an academic author who was trying for the first time to write a book for a broader readership. I will be forever grateful to Louise for giving me this opportunity. As I struggled to organize my thoughts early on, she seemed to understand my message better than I did, which is the highest compliment I can think of giving to a publisher. And her last-minute round of editing provided the icing on the cake.

Jennifer Glossop, my editor, was a delight to work with. She quickly demonstrated complete mastery of the subject matter, always provided me with prompt and insightful constructive feedback, consistently encouraged me throughout the arduous process, and even went so far as to read and comment promptly on draft material – including passages on medical error – while in an intensive care unit watching over her ailing mother. If there's a medal for going above and beyond the call of duty in book editing, Jennifer deserves to receive it and I will enthusiastically nominate her for it.

Although it wasn't their intent, during their lunch together, John Ralston Saul and Jens Rasmussen made me realize just how little I knew about so many important things. They unknowingly inspired me to aim higher and to try to make a difference. Tema Frank encouraged me and gave me invaluable advice when I needed it most. Tad Homer-Dixon and Ken Wiwa also kindly shared their book-writing experiences, providing both insight and encouragement.

Many people read one or more draft chapters: Astrid Otto, Gerard Torenvliet, Jennifer Welsh, Ian Anderson, Caroline Cao, Tema Frank, Nick Dinadis, Kelly Peacock, Paul Cadario, Larry Hettinger, Wendy Trueman, Andrea Cassano, Dave Gaba, Shiphra Ginsburg, Elena Lamberti, Peter Silverman, John Doyle, Asaf Degani, Stu Parsons, Yan Xiao, Bill Rouse, Lianne Jeffs and Barrie Chavel. My parents, André and Marlene, deserve special recognition for reading each and every word of the first draft as it was being produced. Now they know what their son really does for a living – and why. Dianne Howie, Jill Rutherford, Kathleen MacMillan, Asaf Degani, and especially Greg Jamieson and Karima Kada-bekhaled found great source material. Leo Beltracchi and Charles Billings kindly provided details on their work. Wendy Trueman and Julia Glover of CTV gave me permission to use material from their *W-5* program on medical error. Kathleen MacMillan, former Chief Nursing Officer for the Province of Ontario, provided statistics on nurses' overtime and health. Mike Cohen, President of the Institute for Safe Medication Practices, clarified some of the details of the Denver nurses' trial. Stephen Ansolabehere, Professor of Political Science at MIT, shared the initial results of the Cal Tech/ MIT Voting Technology Project. Paul Eisen of CIBC explained some of the limitations of automated phone message systems. Dennis O'Connor, Commissioner of the Walkerton Inquiry, verified the analysis of the Walkerton E. coli outbreak. Klaus Christoffersen, co-author of that analysis and a former student of mine, generously gave me permission to include our joint work here. Doug Harris, Chairman of Anacapa Sciences, Inc., provided useful information about aviation security. Astrid Otto, formerly of Knopf Canada, generously and consistently provided valuable material to help me learn how to express myself more clearly, and did a super job with editorial and administrative responsibilities. Richard Landon, Director of the University of Toronto's Thomas Fisher Rare Book Library, kindly dug up the original printing of Adam Smith's *Wealth of Nations* to see what the typeface looked like. Scott Richardson did a superb job with the design of the cover and the book. Sincere thanks to my graduate students for what they've taught me and for helping me hide periodically during the three years I was working on this book. Release time from teaching and administration so that I could work on the final round of revisions was provided via the generous support of the Jerome Clarke Hunsaker Distinguished Visiting Professorship from the Department of Aeronautics & Astronautics at the Massachusetts Institute of Technology.

Finally, a heartfelt thanks to my parents, to the rest of my family in Canada and Portugal, and to all of my friends – particularly Ian Anderson, Frances Burton, Tema Frank, John Fraser, Margaret Hancock, Chris and Cindy Hunter, Mike Jarcew, Cristina and Gary Kirby, Maurine Kwok, Susan Lang, Patrick LeSage, Suzanne Lethbridge, Elizabeth MacCallum, Ruth Mas, James Orbinski, the late Jane Poulson, John Shaw, Peter Silverman, Rolie Srivastava, Ric Young – and Catherine. The unwavering faith they placed in me prompted me to discover, and believe in, myself.

Index

A&L Canada Laboratories,
  252–254, 263
accountability, desire for, 210–211
activist, 2
advocacy, 238
agricultural chiefdoms, 41
AIDS, 19
air traffic control, 289
aircraft
  B-17, 73, 74
  B-25, 74
  Boeing 727, 195
  BT-13, 73
  C-47, 75
  L-1011, 157
  P-47, 74
airport security, 26, 135–141, 234
  breaches of, 26
  feedback on, 137

  psychological demands of,
    137
  Threat Image Projection
    (TIP) system, 141
  vigilance, 138
Al, Lenore, 255
Albert Einstein College of
  Medicine, 22
Amalberti, René, 294
ambulance dispatching, automatic, 36
amplification effect, 76, 235
anaesthesiologists, anaesthesiology, 151–152,
    169–179, 288
  mortality statistics, 150
anaesthesiology machines
  delivery ports, 151
  lack of delivery port differen-
    tiation, 152
The Andy Griffith Show, 250

Annenberg Center for Health
    Sciences, 22
Apple Newton, 103
apple-and-orange experiment, 47
Applied Ergonomics Corp., 82
Approach Control, 160
The Art of War, 232–233
atomic bomb, 231
Auschwitz, 236
autocratic leader, problems with,
    159
automated phone message sys-
    tem, design of, 90–93
autopilot, 157
aviation, critical incidents, 72
aviation safety, 71–76
    current statistics, 71
    early statistics, 71
Aviation Safety Reporting
        System (ASRS), 216, 249,
        288, 297
    Callback newsletter, 199
    consumers of, 199
    description, 198
    design principles, 200
    history, 196, 197, 198
    as a model for other indus-
        tries, 200
    success of, 202
    use statistics, 201
aviation training. *See* flight crew
    training
Aviation Week Technology
    Innovation Award, 141

"bad apples," 146, 211–216
baggage inspectors, 136
ballot layout. *See* butterfly ballot
    fiasco
batteries, design of, 100
Bechtel, Stone and Webster, 124
Beer, Stafford, 305
behaviour-shaping constraints,
    100, 287
    car keys example, 100
Bell, Dr. Bertrand, 22
Beltracchi, Leo, 128, 131–134
Berra, Yogi, 305
Berwick, Don, 215–216, 219
Beserve, Fred, 114
Bhujwalla, Zoher, 120
bicycles, 240
Billings, Charles, 197–202
birth control pill, 298
Bishop, Ryan, 203–204, 221,
    300
Black & Veetch, 124
"blame and shame," 216, 220,
    222, 224
    harm of, 215
BMW 7 Series. *See* iDrive
bottom-up approach to technol-
    ogy development,
    231–232
boundaries, disciplinary, 30
box cutter, 234
boycotts, 290

Braund, Courtney, 205, 219, 300
Braund, Donnalee, 219, 221
breakfast cereal, 109
breast cancer, 19
Brownie (Eric Clapton's guitar),
        77
Bruce Grey Owen Sound
        Health Unit, 252–253,
        254, 255, 260
Buchanan, Patrick, 281–282
budget allocations, 248
budget cuts, 264–265, 269, 273,
        275–276
Bush, George W., 281–282
butterfly ballot fiasco, 281–285

C-section surgery, 142
Cal Tech. *See* California Institute
        of Technology
California Institute of
        Technology, 283–284
Cape Canaveral, 186
captain, 56, 157–161, 164, 166,
        181
Car and Driver (magazine), 35
Carson, Bill, 78–79
Carter, Jimmy, 304
chads, hanging, dimpled or
        pregnant, 281
Chan, Dr., 169
Chapanis, Lt. Alphonse, 75–76,
        98–99

Chaplin, Charlie, 13, 191
chemical reactor donation, 298
chemotherapy, 203
        errors, 223
Chernobyl nuclear accident,
        9–12, 121, 135
child pornography, 242
Children's Hospital of
        Philadelphia, 222–223
chlorine, chlorinator, 253–254,
        257, 258
        monitors, 256, 262
        residuals, fictitious entries of,
        257
circadian cycle, 127
Clapton, Eric, 77
clocks, VCR and car, 96
CNN, 251
cockpit controls, mismatches,
        73
Cockpit Resource Management
        (CRM), 163–168, 288
        applied to health care, 174.
        See also patient simulator
        cultural effects on, 181
cockpit voice recordings, 164
Cohen, Michael, 212, 214
Combustion Engineering, 124
commercial aviation, accident
        rate, 25
commitment, 277
"Common Sense Revolution,"
        251, 264, 266, 273, 275
        smaller government, 264

communication, 164
    covert, 238–240
complex systems, 20, 24, 25,
    27–28, 34, 38, 60, 110,
    112, 128, 149, 153, 156,
    162–163, 168, 181,
    188–189, 195, 201–202,
    228, 233, 244, 246,
    248–249, 255, 258–260,
    264, 269–277, 288
complex technological systems.
    See complex systems
computerized voting systems,
    283
conflicting objectives, 171
Conservative provincial govern-
    ment. *See* Ontario provin-
    cial government
constraint, behaviour-shaping.
    See behaviour-shaping
    constraints
controlled flight into terrain
    accident. *See* TWA Flight
    514
Cooper, Dr. Jeff, 150–152
coroner's jury, 204
Cowley, Jon, 114
"crazy clocks." *See* VCR auto-
    clocks
crew coordination. *See* team
    coordination
critical incident technique, 72,
    150, 288
curettage, 218

customs interrogation story,
    111
Cyclopean Humanist, 32
Cyclopean Humanistic world
    view, 33, 39, 42, 44,
    46, 48
Cyclopean Mechanist, 32
Cyclopean Mechanistic world
    view, 33, 35, 42, 44, 46,
    48, 73, 202
Cyclops, 32

Darwinian revolution, 305
death by interactions, 212, 272,
    276
debate, 238
defence-in-depth, addition of,
    276
delegation, 159, 164
deliberate harm, 220
Democrats, 281–282
demographic statistics, 235
Denmark, 19, 270, 296
Denver nurses' trial, 208–214
    expert witness testimony,
    212–213
Descartes, René, 30
design changes as discoverable
    evidence, 208
dictatorships, 238
differential equations, 1, 305
dilation, 218

Dilbert, 189, 293
disclosure of errors, positive
    effects of, 226–227
doctors, 23, 43, 59, 149,
    169–170, 172, 176, 179,
    185, 203, 207, 209,
    213–217, 219–220, 225
Doyle, John, 143, 173
drugs
    adverse reactions, 22,
        203–206
    labelling, 204
    reference book, 209
Duchossoir, A. R., 80
DuPont de Neymours & Co.,
    Inc., 82

E. coli. *See* Walkerton E. coli
    outbreak
Eastern Airlines Flight 401. *See*
    Flight 401
Einsteinian revolution, 305
Electric Power Research
    Institute report, 121, 124
electricity
    conservation, 114
    consumption, 27
electrocution, execution by, 87
Elias, Elimson Ribeiro, 172
Emergency Care Research
    Institute (ECRI), 142,
    143, 145

emergent property, 46–49, 271
engineering economic analysis,
    303
engineering education, 189, 302
enthalpy, 129
entropy, 129, 132, 133
environmental problems, 26
equipment failure, 151
ergonomics, 67, 82
European Joint Aviation
    Regulations, 294
execution devices, design of,
    86–88
Experimental Breeder Reactor
    II, 134

FAA-8 test, 137
falsification of records,
    257–258
farming, 40
fascism, 238, 241
Federal Aviation Administration
    (FAA), 136, 141, 196,
    197, 201, 206, 294
feedback, 78–79, 83, 102, 105,
    115, 118, 120, 137, 138,
    199–200, 223, 262, 272,
    275, 277, 292
    bowling analogy, 138
    design principle, 101–102,
    287
Fender, Leo, 78, 79, 80, 81

Fender Stratocaster, 293
  design of, 77–81
fetus, 218
Feynman, Richard P., 188, 268
"Final Solution," 237
financial compensation, 220
first officer, 56, 157–158, 160,
  164, 166, 170
fiscal responsibility, 264
Fish, Kevin, 174
Fitts, Lt. Col. Paul, 72, 73, 74,
  76, 123, 150
Flight 401, 156–159
flight crew, size of, 164
flight crew coordination, effect
  of computer technology
  on, 167–168
flight crew training, 162. *See also*
  Cockpit Resource
  Management (CRM)
flight engineer, 157–158, 160,
  164
flight simulator, fidelity of,
  165–166
Florida, 2000 election outcome,
  281. *See also* butterfly ballot fiasco
Florida Everglades, 157–158
Foley, Patrick, 69, 84, 86
Food and Drug Administration
  (FDA), 144, 145
foreign aid, inappropriately dis-
  pensed, 298
"four-burner" problem, 97–99

Foushee, Clay, 163
Fox television network, 192–193
free speech, lack of, 240

Gaba, David, 174
Gabriel, Peter, 241
genealogical records, 236
General Electric, 123
German government, 298
gestalt, 46
Gibson, James, 48
global population, 27
globalization, 182
Gonzalez, Wayne, 121
Gore, Al, 281–282
government, 260. *See also*
  Ontario provincial gov-
  ernment; Walkerton
  municipal government
Grant, Robert S., 210–211
Green Party, 281
guitar, 37–38. *See also* Fender
  Stratocaster
gypsy, 236

Haitani, Rob, 103, 104, 105
Halden Reactor Project
  Laboratory, 134
handheld computer. *See*
  PalmPilot

Harris, Doug, 137
Harris, Mike, 252, 264, 268
Harrisburg, Pennsylvania. *See*
    Three Mile Island
    nuclear accident
Harvard Medical School, 150
Harvard School of Public
    Health, 146
Hawkins, Jeff, 104
Health Canada, 144, 145
health care, 19–20, 22–24, 27,
    34, 38, 43, 58, 60, 90,
    110, 146–147, 149–152,
    170–171, 173–175, 177,
    179–180, 195, 203,
    206–208, 210, 214–216,
    219, 221–223, 226–227,
    228, 246, 249, 276, 288,
    291–292, 296–297,
    300–301
Helmreich, Robert, 163
Hendrix, Jimi, 81
Hilfiker, Dr. David, 217–220
Hippocrates, 20
Hitler, Adolf, 236
Hollerith, Herman, 235
Hollerith machines, 235–237
homicide, criminally negligent,
    210
homosexual, 236
hopscotch, 109
Howard, Steve, 174
human characteristics, physical,
    65

human error, 17–18, 112, 122,
    142, 151, 191, 223, 270,
    287, 297
    preventible, 147
    so-called, 128, 282
human factors engineering, 1, 2,
    4, 90, 124
human factors engineers, 2–4,
    121, 124, 141
human nature, 45. *See also* tech-
    nology, affinity with
    human nature
human needs and capabilities,
    32
human rights
    abuses, 242
    activists, 241–243
Human-tech, 50–51
Human-tech approach *or* per-
    spective *or* thinking,
    52–54, 62, 65, 286,
    289, 293
    airport security and, 135–141
    aviation and, 75–76, 292
    car key design and, 100
    class project, 113
    cultural effects in team coor-
        dination and, 180
    deception and, 106–108
    Denver nurses' trial and, 211
    developing countries and,
        297
    education and, 302, 304
    execution devices and, 86, 88

guitar design and, 76, 80–81
health care and, 174, 180,
    222–223, 225, 292, 300
human rights and. *See*
    Human-tech approach
    and political activism
lessons of, 278
medical regulations and, 296
nuclear power and, 121–135
PalmPilot and, 106
paper conservation and, 120
patient safety and, 296
PCA device design and, 144
political activism and, 237,
    241, 244
at the political level, 232
power consumption and,
    114
public policy and, 245–246,
    266–267, 269, 295
risk management and, 276
scarcity of, 153
solutions to environmental
    problems and, 113–120
successes, 293
systems management and,
    190
team coordination and,
    162–163
team training and, 179
toothbrush design and, 82
voting technology and, 284
war and, 232
zip-lock bags and, 101

Human-tech ladder, introduced,
    52–62
Human-tech relationships, 60
Human-tech revolution, 4–5,
    51–52, 153, 286, 290,
    292, 296–297, 302, 305
anaesthesia and, 152
aviation and, 76, 289
health care and, 150, 207,
    223, 226
nuclear industry and, 134, 289
at the organization level, 202
role of government in, 295
Human-tech thinkers, attributes
    of, 304
humanist, 2
Humanistic view, 31. *See also*
    Cyclopean Humanist
    world view
hunter-gatherers, 39, 41, 44, 298

IBM, 236–237
iDrive, 35, 44, 51, 104
IEEE Spectrum (magazine), 193
India, 28, 298
Infinia 7220, 35
inquest, coroner's 204
inspections, Ontario Ministry of
    the Environment,
    260–262
Institute for Healthcare
    Improvement, 215

Institute for Safe Medication
    Practices, 143, 212
instructions, importance of, 298
interception rate, 109
International Business Machines.
    See IBM
Internet, use in activism, 243
Internet broadcasting, 242
injection
    intramuscular (IM), 209
    intrathecal (IT), 204–205
    intravenous (IV), 204–205,
    209
invisible hand, 49, 75, 94, 99–100,
    108, 110, 118, 151, 161,
    182, 184–185, 192, 208,
    215, 219, 249, 250, 263,
    266, 273, 276, 278
Ivory Coast, 298

Japan Atomic Energy Research
    Institute, 111
Jew, Jewish, 236–237
Jones, Capt. Richard E., 72, 73,
    74, 123, 150

Kennedy Space Center, 186
Kentucky Veterans'
    Administration Hospital,
    226–227

Kiev, 9
King, Rodney, 242
Kissinger, Henry, 247
Klein, Gary, 140
Klein Associates, 140
Koebel, Frank, 252, 257, 259,
    261, 266, 268–269
Koebel, Stan, 252–261, 263, 266,
    268–269, 272
Koestler, Arthur, 51, 153, 301
Korgaonkar, Dr., 169
Kuk, Dave, 114
Kyoto protocol, 245–246

landing gear
    accidental retraction, 74
    indicator failure, 157
    malfunction, 160
Laplace, Pierre Simon de, 30
lathe. *See* mechanical lathes,
    design of
leadership, 159, 171
Leape, Dr. Lucian, 146–149, 215
learning organizations, 185, 207,
    222
legal liability, as obstactle to
    Human-tech revolutions,
    207–214
legal regulations, 248–249
Lester B. Pearson Airport, 111
lethal injection, execution by, 87
Leuchter, Fred A. Jr., 86–88

leukemia, 203, 205
Liang, Prof. Brian, 208
Lin, Laura, 143
Lockheed Missiles and Space
    Company, 121–124
logic of human destiny, 39, 42, 74
London Ambulance Service, 36
Long, Chris, 120

Machiavelli, Niccoló di
    Bernardo dei, 247
machinists, 69
"Magic Roundabout," 93–95
malpractice lawyers, 227
managerial skills, 189
Manhattan, 234
Manhattan Project, the, 231
Massachusetts General Hospital,
    150
Mayberry, 250
McAuliffe, Christa, 185
McCray, Danielle, 142–143, 145,
    149
McHugh, Prof. Edward, 131
McQuigge, Dr. Murray, 260
mechanical lathes, design of,
    69–70
Mechanistic
    perspective, 68, 70, 191
    thinking, 76, 141, 289
    view, 31. *See also* Cyclopean
    Mechanist world view

medical "class system," 171, 173
medical error, 18, 28, 144,
    146–150, 203, 207–208,
    215–217, 219–222, 224,
    228, 249,
    300, 305
    effects of reporting, 225–227
    psychological effects of, 219
    reporting of, 224–225
    statistics on, 19
    under-reporting of, 206–207
medical fatigue, 21–23
medical residents, 21–22
medication, adverse reaction. *See*
    drugs, adverse reactions
Mercedes-Benz E320, 14, 15
methotrexate, 204–205
Miami International Airport, 157
midget extraterrestrial, 69–70,
    293
miscarriage, mistaken diagnosis
    of, 217
missile defence shield, 301
MIT, 283–284
Modern Times (Charlie Chaplin
    film), 13
monitoring of flight instruments,
    164
morphine overdose, 142
Morris, Errol, 86
Morton Thiokol, 186–187
Moscow, 9

Nader, Ralph, 92, 281

NASA Ames Research Center, 165

Nazi Germany, 236, 237

New England Journal of Medicine, 217

newsletters, covert production of, 241

Norman, D. A. (Don), 90, 290

nuclear power, 121–135

nuclear power plants, poor design of, 122–123

nuclear power plant control room, photo, 122

nuclear reactor. *See* Chernobyl nuclear accident *or* Three-Mile Island nuclear accident

numerical codes, 239

nurse, expemplary record of, 149

nurses, 23, 34, 43, 58, 142–145, 147–149, 169–170, 173–174, 176–177, 207–216, 220, 222–225, 301

O-rings, 186, 187, 188

occupational health and safety, 289

O'Connor, Dennis, 256, 259, 261, 264

O'Hare International Airport, 160

oil check, electronic. *See* Mercedes-Benz E320

one-eyed Humanist *or* Mechanist, 32. *See also* Cyclopean Humanist *or* Mechanist

Ontario provincial government, 251, 255–256, 259, 263–268

Ontario Hospital Association Convention, 300

Ontario Ministry of Health, 252

Ontario Ministry of the Environment, 252, 254, 260, 272

inspections, 261–262

Ontario Provincial Police, 268

operating room

brawl, 169

high tension events, 172

power struggles in, 173

shooting, 172

simulator. *See* patient simulator

organizational decisions

effects of, 188

human nature and, 188

necessity for managers with technical skills, 189–190

Ose, Doug, 139

overtime, nurses and doctors. *See* medical fatigue

Palm Beach County, 282, 283,
    285, 291
PalmPilot, 14, 291, 293
    design of, 103–106
paper
    conservation, 118
    production, 26
paper punch cards, 235
paper voting technology, superi-
    ority of, 284
Parsons, Stuart, 121, 123
patient-controlled analgesia infu-
    sion pump. *See* PCA
    pump
patient safety, 18, 23, 58, 143,
    144, 145–148, 150,
    151–152, 173–174, 185,
    207–208, 211, 214–216,
    223, 225–226, 229, 249,
    296–297
patient simulator
    adoption in medical training,
        179
    details of, 175–176
    effects of described, 177
    experiences with, 178–179
    training curriculum, 177
pattern recognition, 54, 132, 134
PCA pump, 142–150
pedophiles, 242
penicillin, overdose of, 209,
    211–212

perfectibility model for doctors
    and nurses, 147
personal digital assistant (PDA).
    See PalmPilot
pharmacist, 209, 210, 212, 213
phone lines, tapping of, 238
photocopier design for conser-
    vation, 118–120
pilot training. *See* flight crew
    training
pilots, disincentives to report
    near misses, 196–197
Polanyi, John, 301
policy aims, 248–249
policy decisions, 245
policy of extreme honesty, 227
political communication, 241
political decisions as design, 247
politics
    lack of predictability, 246
    role in safeguarding systems,
        265
Portland International Airport,
    160
Portugal, 238
Postman, Neil, 21
Poulson, Jane, 148
Powell, Colin, 301
"Power Pig," 114–118, 287
predictability of politics, 246
pregnancy test, false negative, 217
presidential election, 2000,
    281–285. *See also* butter-
    fly ballot fiasco

privatization of laboratory testing, 263–266
product liability law, 207
protests, World Trade Organization, 242
prototyping, 79, 83, 105
psychomotor skills for flying, 233
public awareness, 276
Public Broadcasting Service (PBS), 192, 193, 194
public census records, 236
public water system, contamination of, 253, 255
punch card identification, 236

quality of life, 3, 14, 18, 24, 26, 40, 48, 51–52, 65, 113, 121, 153, 266, 278, 286, 292, 300

rabbit nets, 39–41
Ramsahai, Deryck, 120
Rasmussen, Prof. Jens, 269–270, 274, 276, 288, 295
Rasmussen risk management framework, 269–278, 288, 295
Reach toothbrush, 245, 293
    design of, 82–84

reading the laundry, 239
Reason, J. (Jim), 270
recounts, hand and machine, 281
Red Tape Commission, 265
reductionism, limitations of, 48
reductionist approach, 29, 46
relationship between people and technology, 49
Republicans, 281
residents. See medical resident
"right stuff," the, 159
Road & Track (magazine), 35
rock 'n' roll, 77

safety, as emergent property, 271
Salazar, Antonio de Oliveira dictatorship of, 238–241
salient body-part hypothesis, 69
saturation temperature, 129
Saul, John Ralston, 249
scapegoats, 214, 225, 263, 269
Schiphol Airport, urinals, 85
Schumacher, Fritz, 299
science versus art, 31
Seattle, 242
secret messages, 239
Seminara, Joseph, 121
September 11th terrorist attacks, 233
shape coding, 76, 287

"shooting the messenger," 57, 196–197
shopping research, 108
Sidney, 205
signal detection theory (SDT), 138–139
silent movies, 13
silos, disciplinary, 30
Silva, Marcelino Pereira da, 172
The Simpsons, 112
sleep deprivation, 21
smallpox, 28
Smith, Adam, 49, 51, 247–248, 267
Snow, C. P., 31
socio-political forces, 246
"soft" technologies, 20, 33, 38–39, 60, 153, 162, 182, 246, 295, 298–299
Space Shuttle *Challenger* accident, 185–188
    contributing organizational factors, 187
special handling, 236
specific volume, 129
splash back and urinal design, 85
Stackhouse, John, 298
steam tables, 129–132, 133
    steps involved in using, 130
stimulus-response compatibility, 287
stimulus-response compatibility principle, 99

stovetop design. *See* four-burner problem
Strat. *See* Fender Stratocaster
student visas, 233
subversive political gathering, 240
Sun Tzu, 232–233
sunk-cost fallacy, 57
superhuman expectation, 39, 148, 150, 152, 184, 215, 220, 223, 228, 263, 267, 295
surgeons, 170–171, 174, 176, 183
Swindon, England. *See* "Magic Roundabout"
Swiss Army knife, 35
syphilis, 209, 212
system, 20
    physical and non-physical aspects, 24
system flaws, Walkerton, 255
systems approach, 49
systems thinking, 46, 48–49, 51

Tallahassee Democrat, 145
Tallahassee Memorial Hospital, 142
Tallahassee state attorney, 145, 148
task analysis, 286

Tavares, Freddie, 78
team coordination
    cultural effects in aviation,
        180–181
    failures in, 156–161,
        169–171
    statistics of failures in, 161
team dynamics, 288
teamwork. *See* team coordina-
    tion
technical skills, 189
technological anthropologists,
    15–16, 24
technology. *See also* "soft" tech-
    nologies
    alienation from, 18
    definition of, 20, 21
    as enabler, 278
    value neutral, 237
    affinity with human nature,
        42, 45, 50, 52, 59, 75, 99,
        143, 153, 156, 162, 191,
        195, 214, 227–229, 246,
        247, 266, 285, 289, 291,
        296, 299, 302, 305
    affinity with human nature,
        test of, 41, 58, 89
terrorism, 136
Thiokol. *See* Morton Thiokol
Third World, 299
Threat Image Projection (TIP)
    system, 141
Three Mile Island nuclear acci-
    dent, 124–128, 296

Thunder Bay, 203–204
Timbuktu, 298
toilet-paper dispenser, design of,
    67–68
toothbrush. *See* Reach tooth-
    brush
top-down approach to tech-
    nology development,
    232
top-down influence, 188
top-down systems design,
    248–250
Toptunov, Leonid. *See* Chernobyl
    nuclear accident
Torres, Charles, 210–213
Toshiba, 17, 35
Toshiba Nuclear Energy
    Laboratory, 134
touch-screen voting systems,
    283, 285
Tracy, Ellen, 222–226
traffic accidents, 19
transitional instability, 40, 42,
    44, 51, 74, 134, 153, 286,
    289, 305
TWA Flight 514, 195–197
two-cultures problem. *See* sci-
    ence versus art

U.S. Bureau of Census, 235
U.S. Environmental Protection
    Agency, 114

U.S. Immigration system, 233
U.S. Institute of Medicine
	(IOM), 18–19, 217
	report, 18–19, 147
U.S. Nuclear Regulatory
	Commission (USNRC),
	124, 128, 132
Uganda, 298
unchlorinated water, effects of,
	261
uncle of the author, 238, 242,
	244
Underhill, Paco, 108
United Airlines, 92
	DC-8-61, 159–160
University of Alberta, 282
urinals
	design of, 84–86
	fly target, 85
	Schiphol Airport, 85
user testing, 104
uterus, 217, 218

varicose veins, surgical near
	miss, 183–184
VCR, 34
	clock, 96
VCR autoclocks, 191–195
	necessity of organizational
		infrastructure in, 194, 195
video cameras, 241
	advantages of, 242

video-conferencing systems,
	56
vigilance, 138
vincristine errors, 203, 205,
	215–216, 219, 221, 300
Vladimir Ilyich Lenin nuclear
	power station. See
	Chernobyl nuclear
	accident
voting, 238
voting technologies, reliability
	of, 283–284

W-5 (news program), 203, 300
Walker, Kristine, 205–206, 221,
	300
Walkerton E. coli outbreak,
	250–269, 296
	boil water advisories, 26,
		255
	public inquiry into, 251
	Well 5, 252, 256–257,
		261–262, 266
	Well 6, 256
	Well 7, 253, 256–257
Walkerton municipal govern-
	ment, 259–261
Walkerton Public Utilities
	Commission (WPUC),
	252–259, 261, 263,
	267–268, 272, 274
waste, 26, 27, 119

water
saturation curve display,
132–134
saturation properties of, 126,
128–129
state of, 126, 132
water operator's certificates, 257
web broadcasts. *See* Internet
broadcasting
Weiner, Earl, 163–164
Westinghouse, 124
Witness (non-governmental
organization), 241–242
Wizards, 33–34, 36–38, 43, 50,
91, 103, 124, 141, 189,
291, 292
work
objectives, 277
schedules, 21
World Trade Center, 234
World Trade Organization
protests, 242
World War II, 72, 231
Wright, Julian, 120
Wright, Robert, 39–42, 44–45,
74, 89, 289
wrong-side surgeries, 184

zero tolerance policy, weaknesses
of, 139
Zion, Libby, 22–23
Zion, Sidney, 22
zip-lock plastic bags, design of,
102, 112

Yarmouth, 205
Yeshiva University, 22

Kim Vicente was Hunsaker Distinguished Visiting Professor of Aeronautics and Astronautics at MIT from 2002 to 2003. In 1999 he was chosen by *TIME* magazine as one of 25 Canadians under the age of 40 as a "Leader for the twenty-first Century who will shape Canada's future." In 2003 he won one of six Steacie Fellowships – a top Canadian science and engineering honour from the Natural Sciences and Engineering Research Council of Canada. He lectures widely around the world and has acted as consultant to, amongst others, NASA, NATO, the US Air Force, the US Navy, Microsoft Corporation and Nortel Networks. Based in Toronto, he is also a professor of engineering at the University of Toronto.

A NOTE ON THE TYPES

The body of *The Human Factor* is set in a digitized form of Caslon, a typeface based on the original 1734 designs of British typefounder William Caslon. It is believed that his original fonts were used in the first edition of Adam Smith's *The Wealth of Nations*, a pioneering work in the field of systems thinking. Throughout its history, Caslon has been widely considered to be one of the most "user-friendly" of all text faces. It is for these reasons that Caslon was selected as the primary face for *The Human Factor*.

The display heads, captions and table/figure texts are set in DIN ("Deutsche Industrie-Norm") Mittelschrift, a typeface originally designed primarily for use on German road signs and license plates. Two of the primary criteria for the DIN type family design are a facility for consistent reproduction, and a clear readability in virtually any point size.